Mobile WiMAX Systems

Performance Analysis of
Fractional Frequency Reuse

Mobile WiMAX Systems

Performance Analysis of Fractional Frequency Reuse

M. K. Salman Alnaimi

Abid Yahya

With contributions from

R. Badlishah Ahmad

CRC Press
Taylor & Francis Group
Boca Raton London New York

CRC Press is an imprint of the
Taylor & Francis Group, an **informa** business
A CHAPMAN & HALL BOOK

CRC Press
Taylor & Francis Group
6000 Broken Sound Parkway NW, Suite 300
Boca Raton, FL 33487-2742

© 2019 by Taylor & Francis Group, LLC
CRC Press is an imprint of Taylor & Francis Group, an Informa business

No claim to original U.S. Government works

Printed on acid-free paper

International Standard Book Number-13: 978-0-8153-4802-3 (Hardback)

Library of Congress Cataloging-in-Publication Data

Names: Fadhil, Mohammed Khalid Salman, author. | Yahya, Abid, author.
Title: Mobile WiMAX systems : performance analysis of fractional frequency reuse / Mohammed Khalid Salman Fadhil, Abid Yahya.
Description: Boca Raton, FL : CRC Press/Taylor & Francis Group, 2019. | "A CRC title, part of the Taylor & Francis imprint, a member of the Taylor & Francis Group, the academic division of T&F Informa plc." | Includes bibliographical references and index.
Identifiers: LCCN 2018014272| ISBN 9780815348023 (hardback : acid-free paper) | ISBN 9781351026628 (ebook)
Subjects: LCSH: Orthogonal frequency division multiplexing. | IEEE 802.16 (Standard) | Mobile communication systems. | Radio resource management (Wireless communications)
Classification: LCC TK5103.484 .F33 2019 | DDC 621.3845/6--dc23
LC record available at https://lccn.loc.gov/2018014272

Visit the Taylor & Francis Web site at
http://www.taylorandfrancis.com

and the CRC Press Web site at
http://www.crcpress.com

This book is gratefully dedicated to my children, my wonderful wife,

and my mother and father, who taught me how to be a man.

M. K. Salman Alnaimi

Dedicated to my family for their love, support, and sacrifice along the path

of my academic pursuits, especially to my father, who took me to school.

Abid Yahya

Contents

List of Figures

List of Tables

Preface

Broadband wireless connections have emerged as a solution to satisfy users' demands for modern E-applications, such as Long Term Evolution (LTE) and Worldwide Interoperability for Microwave Access (WiMAX). According to Ericsson's official projection, global mobile broadband subscriptions will reach 6.5 billion by the end of 2018. This orientation of the public toward wireless broadband Internet services needs more attention in terms of adequate resources and bandwidth to meet users' demands for modern E-services. The world is rapidly moving toward the use of mobile phone applications, and global telecom companies are competing to provide the best services to users. Optimal use of the bandwidth package and available radio resources is the goal of these companies, as they are always working to improve them to match users' requirements. Radio resources are valuable; there is a need for more adaptive and reconfigurable solutions for better exploiting resources in the network, especially when we are talking about the wireless network, since the control signal occupies a significant portion of the bandwidth.

Most of the problems that hinder the best use of radio resources in the cellular network lie in the frequency interference between mobile phone base stations, which results in reduced cell capacity and limits the service quality of cell edge users. Many methods have been proposed to overcome the intercell interference problem. Fractional frequency reuse (FFR) is one of the methods used by WiMAX and LTE technology. FFR is not restricted to WiMAX; it can be used by any mobile technology to address the problem of intercell interference, which is a matter of controlling frequency and time.

An IEEE 802.16e WiMAX base station uses FFR to enhance the capacity and signal quality of cell edge users. Even though FFR is important to use in cellular systems, it has some drawbacks. For this reason, the first chapter of the book explains the principle of intercell interference, the FFR technique, and the problem associated with current FFR. Then Chapter 2 describes the types of FFR, FFR resource assignment challenges in the orthogonal frequency-division multiple access (OFDMA) network, and how to enhance the utilization of radio resources in FFR.

The WiMAX physical layer theory has become an important part of wireless network research, because most of the wireless protocols function there. The most important fetchers of the WiMAX physical layer are related to designing the FFR technique and are presented in Chapter 3, which is very important for understanding, modifying, or enhancing the FFR technique.

Modern mobile communications technology can use two types of frame configuration to match cell load: static and dynamic. In the static configuration, the frame information is changed every several hours or possibly every several days. On the contrary, in the dynamic configuration the frame information is changed frame by frame to match the cell load. Chapter 4 presents the design and implantation of two new FFR algorithms: static resource assignment (SRA) and dynamic resource assignment (DRA). Comprehensive mathematical models are discussed that are not included or clearly explained in other books on WiMAX. Readers can take advantage of these mathematical models in designing, developing, or even applying them to other wireless network systems for different purposes. For future development, this chapter also covers topics like how to effectively model the channel capacity of a cellular network, and the cellular network of 19 base stations.

The results of the proposed algorithms are demonstrated in Chapter 5. In fact, these results represent the outputs of the mathematical models. The chapter shows the effort of the new FFR algorithms in the cellular network of 19 base stations. Also, these algorithms are tested by applying three types of WiMAX system parameters, such as 5, 10, and 20 MHz system bandwidth.

Acknowledgments

We thank the following individuals; without their contributions and support, this book could not have been written.

We would like to express their special gratitude and thanks to the Ministry of Higher Education and Scientific Research and Science and Technology, Iraq; Botswana International University of Science and Technology (BIUST); and Universiti Malaysia Perlis (UniMAP), Sarhad University of Science and Information Technology (SUIT), Peshawar, for giving us the attention, time, and opportunity to publish this book.

We would also like to take this opportunity to express our gratitude to all those people who provided us with invaluable help in the writing and publication of this book.

About This Book

In this book, two different algorithms are designed, modeled, and implemented to enhance the capacity, coverage area, and radio resource utilization of WiMAX IEEE 802.16e cellular networks using fractional frequency reuse (FFR).

Two types of frame configuration are considered to match cell load: static and dynamic. In the static configuration, the frame information is changed every several hours or possibly every several days.

On the contrary, in the dynamic configuration the frame information is chanced frame by frame to match the cell load. Each method can be used for specific E-applications. Whereas in the static algorithm five cases are applied to choose the best configuration, in dynamic algorithm two types of mobility pattern are applied to tackle the variation in population density. These algorithms represent different solutions to handle different user types.

This book presents comprehensive mathematical models that are not included or clearly discussed in other WiMAX books. Readers can take advantage of these mathematical models by designing, developing, or even applying them to other wireless network systems for different purposes. Furthermore, channel capacity modeling for three types of WiMAX system profiles (5, 10, and 20 MHz) is mathematically presented.

Authors

M. K. Salman Alnaimi earned his BSc degree in electrical and electronic engineering from the University of Technology, Iraq. He earned his MSc and PhD degrees in computer and communication engineering from the School of Engineering, University of Technology, and the University Malaysia Perlis, respectively. He has worked in the Ministry of Science and Technology as the director of the Digital Society Research Center, and he was also the project manager of an E-ministry project (enterprise level). His research interests focus on wireless communication with an emphasis on mobile WiMAX networks, network deployment, and information security. Recently, he has become interested in E-governance.

Abid Yahya began his career on an engineering path, which is rare among other researcher executives, and he earned his MSc and PhD degrees in wireless and mobile systems from the Universiti Sains Malaysia, Malaysia. Currently, he is working at the Botswana International University of Science and Technology. He has applied this combination of practical and academic experience to a variety of consultancies for major corporations.

He has more than 115 research publications to his credit in numerous reputable journals, conference articles, and book chapters. He has received several awards and grants from various funding agencies and supervised a number of PhD and master candidates. His new book, *LTE-A Cellular Networks: Multi-Hop Relay for Coverage, Capacity and Performance Enhancement*, was published by Springer International Publishing in January 2017 and is being followed in national and international universities.

Professor Yahya was assigned to be an external and internal examiner for postgraduate students. He has been invited a number of times to be a speaker or visiting lecturer at different multinational companies. He sits on various panels with the government and other industry-related panels of study.

List of Symbols

R1 zone	First part in the DL subframe
R3 zone	Second part in the DL subframe
F1	Frequency band of segment A
F2	Frequency band of segment B
F3	Frequency band of segment C
N_{FFT}	Number of subcarriers
Δf	Subcarrier frequency spacing
Tb	Useful symbol time
Tg	Guard time (or cyclic prefix time)
G	Ratio of cyclic prefix time to useful symbol time
Ts	OFDMA symbol duration time
n	Sampling factor
N_{bin}	Number of bins per slot
M_{OFDM}	Number of OFDM symbols per slot
T1	Operational time of the R1 zone
T2	Operational time of the R3 zone
N_{MS}	Number of users requiring services
$SINR_{TH}$	Threshold SINR
$SINR(MS)$	Mobile station SINR value
r	Cell center radius
R	Cell radius
No	Thermal noise power
l	Number of interfering base stations in the grid
M	Target zone or segment name
Trd. FFR	Traditional fractional frequency reuse
Pro. FFR	Proposed fractional frequency reuse
Seg. BC	Segment BC
S_{frame}	Number of OFDM symbols in WiMAX frame
Tf	Frame duration time
Kr_{slot}	Number of subcarriers per slot
N_{Smb}^{UL}	Number of OFDM symbols in the UL subframe
DL/UL	Downlink-to-uplink OFDM traffic ratio
N_{MS}^{Gen}	Number of generated users
$N_{MS_{R1}}^{max}$	Maximum number of users in the R1 zone
N_{MSR1}	Number of users in the R1 zone
$N_{MS_{R3}}^{max}$	Maximum number of users in the R3 zone
N_{MSR3}	Number of users in the R3 zone
N_{MS}^{Extra}	Number of extra users requiring services
MS_x	User location in the X-axis

MS_y	User location in the Y-axis
$ic1, ic2, ic3,$ and $ic4$	User counters of cases 1, 2, 3, and 4, respectively
$ic1^{max}, ic2^{max}$ $ic3^{max}, ic4^{max}$	Maximum number of users that can be served in cases 1, 2, 3, and 4, respectively
N_{R1}	User counter of the R1 zone
$N_{MS}^{LA}, N_{MS}^{LB}, N_{MS}^{LC}, N_{MS}^{LD}$	Number of users per layers A, B, C, and D, respectively
MSF	User index flag
N_{MS}^{ABCD}	Total number of users in all layers (A, B, C, and D)
N_{MS}^{max}	Holds the most crowded layer name (A, B, C, or D)
ia, ib, ic, id	User counters of layers A, B, C, and D, respectively
$ia^{max}, ib^{max}, ic^{max}, id^{max}$	Maximum number of users in layers A, B, C, and D, respectively
Pr	Received power
Pt	Transmitted power
Gr	Receiver antenna gain
Gt	Transmitter antenna gain
f	Operating carrier frequency
h_{BS}	Base station antenna height
h_{MS}	Mobile station antenna height
$A(h_{MS})$	Mobile station antenna correction factor
d	Distance between base station and user (or mobile station)
C_F	Environment correction factor
X	Shadowing
BS_x	Base station location in the X-axis
BS_y	Base station location in the Y-axis
K	Boltzmann's constant
T	Kelvin temperature
Fs	Sampling frequency
Kr_{OFDM}	Number of subcarriers per OFDM symbol
b	Number of data bits per subcarrier
Dr_{PHY}	Physical layer data rate
Dr_{MAC}	MAC layer data rate
N_{OFDM}^{DL}	Number of OFDM symbols in the DL subframe
N_{OFDM}^{OH}	Number of OFDM symbols reserved for overhead (control messages) in the DL subframe
Cr	Code rate type
Q	Number of points in the constellation for a particular modulation type

P(u)	Binary expression (0 or 1)
$P_{\text{SINR}}^{\min}(M)$ and $P_{\text{SINR}}^{\max}(M)$	SINR thresholds for target zone or segment
$R3_A$	Segment A in the R3 zone
$R3_{BC}$	Segment BC in the R3 zone
$Dr_{\text{MAC}}^{\text{Trd}}$	MAC data rate in traditional FFR
$Dr_{\text{MAC}}^{\text{Pro}}$	MAC data rate in proposed FFR
Z	Number of trials
$\overline{Dr}_{\text{MAC}}^{\text{Trd}}$	Average MAC data rate in traditional FFR
$\overline{Dr}_{\text{MAC}}^{\text{Pro}}$	Average MAC data rate in proposed FFR
$Kr_E(M)$	Normalized subcarrier efficiency per zone or segment
$\overline{Kr}_E(M)$	Average normalized subcarrier efficiency per zone or segment
Kr_E^{Trd}	Arithmetic mean of traditional FFR subcarrier efficiency
Kr_E^{Pro}	Arithmetic mean of proposed FFR subcarrier efficiency
DL_{SE}	Downlink spectral efficiency
FRF_{R3}	Frequency reuse factor in the R3 zone
$FRF_{R1}e$	Frequency reuse factor in the R1 zone
dr_{DL}	Data rate per user in the DL subframe
N_{DL}	Number of active users in the DL subframe
$N1$	Number of OFDM symbols in the R1 zone
$N3$	Number of OFDM symbols in the R3 zone
Nt	Number of OFDM symbols in the R1 and R3 zones
$\overline{DL_{SE}}$	Average DL spectral efficiency
$N_{\text{slot}}(M)$	Number of utilized slots per zone or segment
S_{burst}^M	Number of slots per burst per zone or segment
$N_{\text{subch}}(M)$	Number of used subchannels in particular zone or segment
$N_{\text{OFDM}}(m)$	Number of OFDM symbols in other parts of the DL subframe that are not equal to the current active part
$N_{\text{slot}}^{\text{Trd}}$	Total number of used slots in traditional FFR
$N_{\text{Slot}}^{\text{Pro}}$	Total number of used slots in proposed FFR
$\overline{N}_{\text{slot}}^{\text{Trd}}$	Average number of used slots in traditional FFR
$\overline{N}_{\text{slot}}^{\text{Pro}}$	Average number of used slots in proposed FFR
$N_{\text{user}}(M)$	Number of active users in the intended zone or segment

User(u)	A user indexed
$N_{\text{slot}}^{\max(M)}$	Maximum number of slots specified for a certain zone or segment
$N_{\text{user}}^{\text{Trd}}$	Total number of active users in traditional FFR
$N_{\text{user}}^{\text{Pro}}$	Total number of active users in proposed FFR
$\bar{N}_{\text{user}}^{\text{Trd}}$	Average number of active users in traditional FFR
$\bar{N}_{\text{user}}^{\text{Pro}}$	Average number of active users in proposed FFR
C	Channel capacity
C_{FFR}	Channel capacity in FFR technique
N_{subch}	Total number of available subchannels in the system
$C_{\text{FFR}}^{\text{Trd}}$	Channel capacity in traditional FFR
$C_{\text{FFR}}^{\text{Pro}}$	Channel capacity in proposed FFR
DRA-I	Dynamic resource assignment algorithm using mobility pattern I
DRA-II	Dynamic resource assignment algorithm using mobility pattern II
SRA FFR Cs. 1	Static resource assignment algorithm using case 1
SRA FFR Cs. 3	Static resource assignment algorithm using case 3
Seg. BC-I	Response of segment BC in mobility pattern I
Seg. BC-II	Response of segment BC in mobility pattern II
Pro. FFR-BC	Channel capacity of the proposed SRA and DRA algorithms when only subchannels of segment BC are considered
Pro. FFR-ABC	Channel capacity of the proposed SRA and DRA algorithms when all of the available subchannels in segments A, B, and C are considered
α	Number of active users in the target zone or segment
β	Number of slots reserved for user data load
γ	Number of slots per two successive OFDM symbols
δ	Ratio of used bandwidth
ρ	Ratio of data OFDM symbols used by a particular zone or segment to the total number of data OFDM symbols occupied in the DL subframe
τ	Operational time of a given OFDM symbol in the DL subframe
φ	Required number of OFDM symbols per slot
ω	Total number of subcarriers reserved for a specific user load

List of Abbreviations

3GPP	Third-Generation Partnership Project
AAS	Adaptive antenna system
AFRF	Average frequency reuse factor
AMC	Adaptive modulation and coding
AP	Access point
APA	adaptive power allocation
AT&T	American Telephone and Telegraph Company
Band-AMC	Band adaptive modulation and coding
BE	Best effort
BP	Burst profile
BS	Base station
BW	Bandwidth
BWE	Bandwidth efficiency
CC	Convolutional coding
CC-FFR	Client-centric fractional frequency reuse
CCI	Co-channel interference
CDMA	Code-division multiple access
COST	European Cooperative for Scientific and Technical
CQI	Channel quality indicator
CSI	Channel state information
CTC	Convolutional turbo code
DC	Direct current
DIUC	Downlink Interval Usage Code
DL	Downlink
DL-MAP	Downlink MAP
DRA	Dynamic resource assignment
DSA	Dynamic subcarrier assignment
DSL	Digital subscriber line
E	Electronic
EBW	Effective bandwidth
ertPS	Extended Real-Time Polling Service
FCH	Frame control header
FDD	Frequency-division duplexing
FEC	Forward error correction
FFR	Fractional frequency reuse
FFT	Fast Fourier transform
FRF	Frequency reuse factor
FTP	File Transfer Protocol
FUSC	Full Usage of Subchannels
GSM	Global System for Mobile

H	High
HS	Highest
HTTP	Hypertext Transfer Protocol
ICI	Intercell interference
IE	Information element
IEEE	Institute of Electrical and Electronics Engineers
IFCO	Interference coordination
ISD	Intercell spatial demultiplexing
ISI	Intersymbol interference
L	Low
LAN	Local area network
LMSC	LAN MAN Standards Committee
LOS	Line-of-sight
LS	Lowest
LTE	Long Term Evolution
MAC	Media access control
MAN	Metropolitan area network
MCS	Modulation and Coding Scheme
MPEG	Moving Pictures Experts Group
MRC	Maximal ratio combining
MS	Mobile station
MSINR	Maximum SINR
NLOS	Non-line-of-sight
nrtPS	Non-Real-Time Polling Service
OFDM	Orthogonal frequency-division multiplexing
OFDMA	Orthogonal frequency-division multiple access
PHY	Physical layer
PL	Path loss
PMP	Point-to-multipoint
PUSC	Partial Usage of Subchannels
QAM	Quadrature amplitude modulation
QoS	Quality of service
QPSK	Quadrature phase-shift keying
RD	Radio distance
REP-REQ	Report request
REP-RSP	Report response
RNC	Radio network controller
RR	Round-robin
RRA	Radio resource agent
RRC	Radio resource controller
RSSI	Received signal strength indicator
RTG	Receive transition gap
rtPS	Real-Time Polling Service
SE	Spectral efficiency
Segment BC	Segments B and C in the R3 zone

SINR	Signal-to-interference-plus-noise ratio
SNR	Signal-to-noise ratio
SRA	Static resource assignment
TDD	Time-division duplexing
TLPC	Two-level power control
TTG	Transmit transition gap
U	User
UGS	Unsolicited Grant Service
UIUC	Uplink Interval Usage Code
UL	Uplink
UL-MAP	Uplink MAP
VOIP	Voice over Internet Protocol
WiMAX	Worldwide Interoperability for Microwave Access

Basic Units

Quantity	Unit	Symbol
Data rate	Bits per second, megabits per second	bps, Mbps
Frequency	Hertz, kilohertz, megahertz, gigahertz	Hz, kHz, MHz, GHz
Distance	Meter, kilometer	m, km
Time	Second, millisecond, microsecond	s, ms, μs
Spectral efficiency	Bits per second per hertz	bps/Hz
Subcarrier efficiency	Bits per subcarrier per burst	b/subcarrier/burst
Channel capacity	Bits per second	bps
Power	Watt, milliwatt, decibel, decibel-milliwatt	W, mW, dB, dBm
Antenna gain	Decibel-isotropic	dBi
Thermal noise power	Decibel-milliwatt per hertz	dBm/Hz

1

Introduction

1.1 Background

In recent years, there has been a great deal of interest in E-applications to meet the requirements of different segments of the public, such as E-commerce, E-health, E-business, and E-learning. Today, access to these applications is not limited to fixed terminals (computers); instead, it mainly depends on mobile devices, such as cellular phones. In a cellular system using point-to-multipoint topology, the base station is the communications coordinator [1]. Therefore, these base stations must be highly efficient in order to satisfy users' demands. However, wired broadband communication offers higher-data-rate access to the Internet, such as the fixed digital subscriber line (DSL) [2]. Hence, it requires a more costly infrastructure than does wireless communication. In contrast, wireless communications solve the problem of the high cost of infrastructure and facilitate access to E-applications, regardless of the location of the users.

Worldwide Interoperability for Microwave Access (WiMAX) is a broadband wireless technology that has emerged as an alternative to DSL communication and brings the broadband line to the air. WiMAX base stations can be efficiently used in a cellular system, since they use orthogonal frequency-division multiple access (OFDMA) based on orthogonal frequency-division multiplexing (OFDM). Nevertheless, in the cellular system, intercell interference (ICI) is a major problem because of the subcarriers' collision. ICI occurs as a result of using the same subcarriers in neighboring base stations at the same time. The subcarriers in the OFDM technique are orthogonal; the orthogonality solves the problem of intercarrier interference, and, adding a guard interval to the OFDM symbol, solves the problem of intersymbol interference (ISI) as well. The ICI has nothing to do with intercarrier interference or ISI. The OFDMA is a multiple access technique that helps divide the available subcarriers into groups of virtual subchannels. These subcarriers can be allocated randomly per group of subchannels to avoid collisions, which mitigate the ICI effect. Figure 1.1 shows an example of subcarrier collision in the cell border of a cellular system using

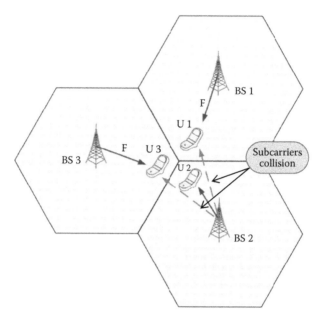

FIGURE 1.1
Interference model in a cellular system using universal frequency.

a frequency reuse factor (FRF) of 1 [3], where all the available bandwidth is reused by every cell in the grid.

As can be seen in Figure 1.1, three mobile stations operate on the same frequency band (F). For instance, base station (BS) 2 communicates with user (U) 2 using the same frequency band (F) as BS1 and BS3, which results in a source of interference to other mobile stations in the network (U1 and U3). The ICI greatly affects the channel quality of cell border users.

The random distribution of subcarriers reduces the chance of subcarrier collision. However, an increase in the number of active users increases the probability of collision. As a result, network designers have found different ways to reduce the ICI effect, such as conventional frequency planning, sectoring, and fractional frequency reuse (FFR). Figure 1.2 shows an example of the FFR technique used in cellular network deployment to reduce the ICI in the cell border [4].

OFDMA facilitates the allocation of user resources in the time domain (slots) and frequency domain (subchannels); therefore, the allocation of radio resources can be controlled in time and frequency for every user. In order to achieve the goal of the FFR technique, the downlink (DL) subframe is divided into two parts. The first one is called the R1 zone and is used to serve users of a cell center area (yellow part), where all the

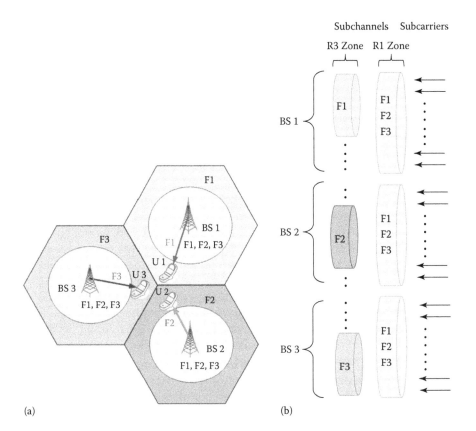

FIGURE 1.2
FFR technique example. (a) Cell border frequency distribution. (b) Subcarrier grouping in subchannels.

available subchannels (bandwidth) are used (F1, F2, and F3), as shown in Figure 1.2b. The second part is called the R3 zone and is used to serve users of a cell border area. In the R3 zone, the subchannels (bandwidth) are divided into three segments or three bands (F1, F2, and F3), where each cell can only use one band to serve cell border users, as illustrated in Figure 1.2a.

FFR uses one-third of the available subchannels to serve cell border users (R3 zone) [5]. Thus, the ICI is mitigated since different subcarriers are used in each cell border at the same time slot. However, using one-third of the available subchannels (or bandwidth) in each cell to serve cell border users leads to inefficient use of resources and bandwidth (radio resources) [6]. The inefficient utilization of radio resources reduces the number of served users and data rate, as well as spectral efficiency and channel capacity.

1.2 FFR Implementation Complexity

This book has identified the following problems in the present FFR technology:

1. The FFR technique aims to enhance the signal quality of cell border users by using one-third of the available bandwidth in each cell edge of the grid. However, this enhancement of cell border signal quality is at the cost of losing two-thirds of the available radio resources in each DL subframe of each base station [4]. Consequently, these losses in radio resources lead to shortage in resources, data rate, and spectral efficiency, as well as reducing the number of served users and channel capacity. Most of the current research does not focus on the loss of two-thirds of the radio resources. This book focuses on enhancing the performance of the FFR technique by different trends, such as intelligently allocating the subcarrier to reduce the interference [7], selecting the zone of operation (R1 or R3) [8] while maintaining the quality of service requirements in order to maximize the resource utilization [9], and considering the free slots in the DL subframe when making decision to assign resources to users [10].

2. The distribution of mobile users in the cell coverage area is random and variable due to the movement of these users from one place to another, which leads to a variation in population density. To address this variation in population density, the dynamic configuration method should be used to update the system parameters frame by frame according to cell load, such as the size of the R1 and R3 zones. The dynamic configuration enhances the base station performance; however, the DL subframes need to be continuously reconstructed (the zone sizes are automatically resized), which increases FFR implementation complexity [11,12].

1.3 WiMAX Capacity-Based Cellular System

The scope of this book focuses on algorithms to improve the WiMAX capacity-based cellular system, where two new algorithms are proposed: static resource assignment (SRA) FFR and dynamic resource assignment (DRA) FFR. These two algorithms modify and enhance the operation of the FFR technique by using all the available bandwidth (subchannels) to serve extra users without increasing the interference level and system cost.

SRA enhances the performance of the FFR technique in terms of data rate, spectral efficiency, resource utilization, and number of served users, where all of the available bandwidth is used [6,13]. This means that a frequency reuse of 1 is achieved. The SRA is presented on the basis of the static configuration method and does not involve more partitioning of the DL subframe instead of exploiting the untapped space (slots) in the DL subframe. Four cases are proposed to investigate best SRA performance. These cases help us find the best configuration of SRA. However, to address changes in population density, a dynamic DRA FFR is presented as well [14]. The dynamic method presented herein through two mobility patterns shows the benefits of the dynamic model.

References

1. L. Nuaymi, *WiMAX: Technology for Broadband Wireless Access*, Wiley, Hoboken, NJ, 2007.
2. Y. Zhang and H.-H. Chen, *Mobile WiMAX: Toward Broadband Wireless Metropolitan Area Networks*, CRC Press, Boca Raton, FL, 2007.
3. J. G. Andrews, A. Ghosh, and R. Muhamed, *Fundamentals of WiMAX: Understanding Broadband Wireless Networking*, Pearson Education, London, 2007.
4. Forum, Mobile WiMAX—Part I: A technical overview and performance evaluation, 2006, p. 53.
5. M. Salman, R. Ahmad, Z. G. Ali, J. A. Aldhaibani, and R. A. Fayadh, Analyzing mobile WiMAX base station deployment under different frequency planning strategies, in *International Conference on Mathematics, Engineering and Industrial Applications, (ICoMEIA) Malyasia, 2014*, 2015, p. 070072.
6. M. Salman, R. Ahmad, and A. Yahya, A new approach for efficient utilization of resources in WiMAX cellular networks, *Tehnicki vjesnik/Technical Gazette*, vol. 21, pp. 1385–1393 2014.
7. Z. Mohades, V. T. Vakili, S. M. Razavizadeh, and D. Abbasi-Moghadam, Dynamic fractional frequency reuse (DFFR) with AMC and random access in WiMAX system, *Wireless Personal Communications*, vol. 68, pp. 1871–1881, 2013.
8. B. Pijcke, M. Gazalet, M. Zwingelstein-Colin, and F.-X. Coudoux, An accurate performance analysis of an FFR scheme in the downlink of cellular systems under large-shadowing effect, *EURASIP Journal on Wireless Communications and Networking*, vol. 2013, pp. 1–14, 2013.
9. M. Einhaus, A. Mäder, and X. Pérez-Costa, A zone assignment algorithm for fractional frequency reuse in mobile WiMAX networks, in *Networking 2010*, Springer, Berlin, 2010, pp. 174–185.
10. I. N. Stiakogiannakis, G. E. Athanasiadou, G. V. Tsoulos, and D. I. Kaklamani, Performance analysis of fractional frequency reuse for multi-cell WiMAX networks based on site-specific propagation modeling [wireless corner], *IEEE Antennas and Propagation Magazine*, vol. 54, pp. 214–226, 2012.
11. S.-E. Elayoubi, O. Ben Haddada, and B. Fourestie, Performance evaluation of frequency planning schemes in OFDMA-based networks, *IEEE Transactions on Wireless Communications*, vol. 7, pp. 1623–1633, 2008.

12. L. Li, D. Liang, and W. Wang, A novel semi-dynamic inter-cell interference coordination scheme based on user grouping, in *IEEE 70th Vehicular Technology Conference Fall (VTC 2009–Fall)*, Anchorage, AK, 2009, pp. 1–5.
13. M. Salman, R. Ahmad, and M. S. Al-Janabi, A new static resource and bandwidth utilization approach using WiMAX 802.16 e fractional frequency reuse base station, *Journal of Theoretical and Applied Information Technology*, vol. 70, 2014, pp. 140–152.
14. M. Salman, B. Ahmad, and A. Yahya, New dynamic resource utilization technique based on fractional frequency reuse, *Wireless Personal Communications*, vol. 83, pp. 1183–1202, 2015.

2

FFR WiMAX Base Station Deployment

2.1 Introduction

WiMAX is a broadband evolution in a wireless context and is a wireless communication technology for delivering high-speed connections. Computers and E-applications have become essential in our lives. The Internet is the platform that provides support for business, finance, learning, playing, and so on. WiMAX is a promising technology for accessing the Internet infrastructure and information highway.

The standard for WiMAX was developed by the Institute of Electrical and Electronics Engineers (IEEE) in sequence phases, all started by the IEEE-802 LAN/MAN Standards Committee (LMSC), which was formed in February 1980 to define standards for local and metropolitan area networks (LANs and MANs) [1]. The first release of 802.16 WiMAX was in October 2001, and it was published on April 8, 2002 [2]. It can be considered an alternative to cable networks such as digital subscriber line (DSL), cable modem, and fiber optic, where a physical connection is required. However, the line-of-sight (LOS) condition is assumed for this standard. The non-line-of-sight (NLOS) connection standard IEEE 802.16-2004 emerged in 2004 for frequencies below 11 GHz [3]. This standard specifies orthogonal frequency-division multiplexing (WirelessMAN OFDM) transmission technology with a carrier frequency of 3.5 GHz and is denoted as fixed WiMAX. In 2005, the IEEE 802.16e standard was approved, and it was published in 2006. It uses an orthogonal frequency-division multiple access (WirelessMAN OFDMA) physical layer with a carrier frequency of 2.5 GHz and is denoted as mobile WiMAX. It can be considered an amendment to the fixed WiMAX 802.16–2004 [4]. This chapter surveys related literature on fractional frequency reuse (FFR) WiMAX base station technology. It is arranged as follows: Section 2.2 describes the cellular network concept. Section 2.3 explains the intercell interference (ICI) phenomenon that occurs in cellular network deployment. Section 2.4 describes the frequency planning and FFR as basic techniques that are used to address the ICI in cellular networks. Section 2.5 introduces the classification of the FFR technique, while Section 2.6 highlights the challenges and possible solutions to improve the performance of the FFR technique. Finally, Section 2.7 presents a summary of this chapter.

2.2 Cellular Network Systems

The available spectrums in the market are scarce compared with the tremendous growth in the demand for increased communication services, which definitely leads to the restructuring of mobile networks. Network capacity and coverage area are the main concerns of cellular network designers. The first solution to increase the network capacity was proposed by researchers of American Telephone and Telegraph Company (AT&T) Bell Laboratories, where the concept of cellular networks was first introduced [5].

The cellular network concept overcomes spectral congestion and a reduction in cell capacity, where it enables the base station to reuse part of the available spectrum without a need for significant changes in technology. The cellular system replaces a high-power transmitter to cover a large cell area with many low-power transmitters to cover small areas; thus, the capacity is increased and the spectrum reuse in group of cells became possible [6]. An example of the cellular system is shown in Figure 2.1 [7].

In the cellular network shown in Figure 2.1, the cluster size equals seven cells, assuming that each cell operates on a different frequency band. As a result, the frequency reuse factor (FRF) equals 7. The FRF represents the number of frequency bands that can be used by a given number of cells; for instance, if FRF = 3, then the frequency bands are reused every three cells [7]. The ICI in each cluster is mitigated, but the same frequency bands are reused in the neighboring clusters, which results in interference between cells. As illustrated in Figure 2.1, the mobile user in cell 1 of cluster A is affected by the co-channel, where the same frequency band is reused at cell 1 of cluster B. The blue arrow represents the frequency reuse distance, which should be far

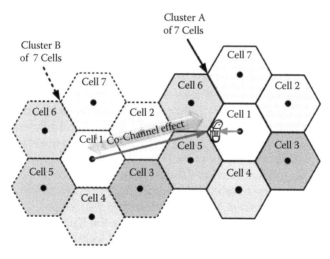

FIGURE 2.1
ICI in cellular network deployment.

enough to reduce the ICI and keep the signal-to-interference-plus-noise ratio (SINR) at an acceptable level, especially at the cell border. Basically, the blue arrow represents an undesired signal received by the mobile user, whereas the green arrow represents the desirable signal received from its base station. The co-channel effect (i.e., ICI) causes loss of service, a reduction in the throughput, shrinkage of the cell coverage area, and an increase in the bit error rate [6].

2.3 Intercell Interference Mitigation

ICI is a major problem in OFDMA cellular systems like WiMAX and Third-Generation Partnership Project (3GPP) Long Term Evolution (LTE). Many solutions have been proposed to solve this problem, such as topology control [8], sectorization, smart antenna [9], interference averaging, and interference avoiding [10,11].

The 3GPP LTE introduces the interference coordination (IFCO) technique to solve the ICI problem, which requires the base stations to coordinate their transmissions [12]. The IFCO is classified based on the time level of operation in three techniques: static, semistatic, and dynamic. In the static IFCO technique, system parameters are set during the design stage, and it operates on a timescale of several days. The semistatic technique considers the variation in cell load and user load and operates on a timescale of several minutes. Finally, in the full dynamic IFCO technique, the system parameters are changed dynamically according to the variation of traffic load or users' distribution. Therefore, it operates on a timescale in the order of single frame.

In WiMAX, two main techniques were suggested by [11] to mitigate the ICI effects, namely, interference avoidance and interference averaging. Interference avoidance involves frequency planning where the network bandwidth is divided among cells or sectors in a certain way to ensure that each cell or sector operates with a different frequency band. In contrast, interference averaging is based on the fact that the subcarrier's mapping into the subchannels is a random process (pseudorandom) that reduces the chance of using the same subcarrier in the neighboring cells or sectors (collision); thus, interference will be averaged out between all users [13]. The interference avoidance requires carefully dividing the available bandwidth between cells or sectors (optimum frequency planning) to avoid wasting radio resources. On the other hand, in interference averaging, subcarrier allocation is changed randomly at every opportunity of transmission. Therefore, interference averaging requires strong design tools to ensure best network performance. These two techniques can be used together to control the ICI effect in the cellular system.

2.4 Base Station Deployment in OFDM-Based Cellular Network

Base stations can be deployed adjacent to each other in a cellular network to cover a wide area. Network operators try to divide the available bandwidth (or spectrum) among acceptable number of cells in order to efficiently utilize the bandwidth and reduce the effect of ICI as much as possible. The main advantages that can be obtained when reducing the ICI are an improvement of the cell coverage area and capacity. In the subsequent sections, two techniques are discussed to address the ICI problem: frequency planning and FFR.

2.4.1 Frequency Planning Technique

The frequency planning technique is the process of dividing the available bandwidth among cells or sectors in order to reduce the interference between cells. It is possible to implement many frequency planning schemes in an OFDMA cellular network; three of these schemes are depicted in Figure 2.2 [11]. When FRF equals 1, as in Figure 2.2a [14], high throughput

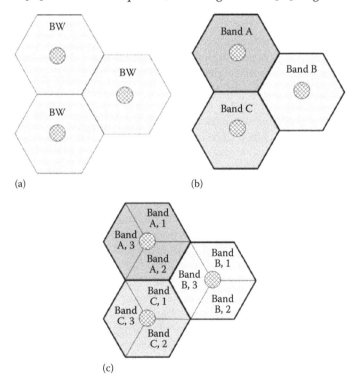

FIGURE 2.2
Three types of frequency planning schemes: (a) FRF of 1, (b) FRF of 3, and (c) sectoring.

can be achieved because all of the available bandwidth is used by each cell. However, users at the cell edge suffer from high interference due to the fact that the same frequency bandwidth is reused in the neighboring cells (co-channel effect), and therefore the interference averaging technique is required in such a case.

In Figure 2.2b, the FRF is equal to 3 [14], where each cell operates with part (one-third) of the bandwidth. This results in a mitigation of the interference between the adjacent cells. The ICI is reduced, but at the cost of losing two-thirds of all available bandwidth in each cell. In the same context, if sectoring is applied as in Figure 2.2c [14], the impact of interference is less, but again, the bandwidth is divided by 3. Nevertheless, with sectoring the SINR is increased, which means that a high modulation order can be used, which ends up compensating for sacrificing the bandwidth. As a result, the frequency planning technique can be used to mitigate the ICI effect, but at the expense of not using all the available bandwidth.

2.4.2 Fractional Frequency Reuse

The FFR technique is widely used in cellular network deployment. The concept was first applied in Global System for Mobile (GSM) networks [15] and then adapted in the WiMAX Forum [16] and LTE [17]. In the FFR technique, the available resources are divided into two zones, R1 and R3 zones, and the cell area is also divided into two virtual regions: the cell center region and the cell border region, as illustrated in Figure 2.3 [16].

The FFR aims to avoid ICI by using a fraction of the available frequency band (or bandwidth) in the cell border. Segment A in the R3 zone is used to provide one-third of the available bandwidth to serve users at the cell border. Conversely, in the R1 zone all the available bandwidth is used to serve users

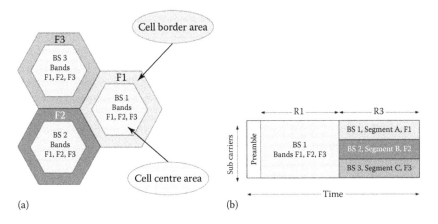

(a) (b)

FIGURE 2.3
Deployment of three FFR base stations: (a) frequency distributions and (b) DL subframe structure.

in the cell center. However, the available bandwidth in the R3 zone is divided into three equal and orthogonal frequency bands (or segments) named F1, F2, and F3. These bands can be used in the cell border of adjacent cells to reduce the interference between these cells.

In Figure 2.3b, the downlink (DL) subframe structure is illustrated, and in Figure 2.3a, the virtual frequency distributions of three base stations are depicted. In order to attain the benefit of using FFR, only one segment out of three should be used at the R3 zone in each base station. However, this sacrifice in resources and bandwidth is mandatory to reduce the ICI at the cell edge [18]. The Partial Usage of Subchannels (PUSC) permutation mode should be used to divide the R3 zone into three segments [16]. Accordingly, the main advantage of using the FFR technique is to increase the signal strength of the users, especially at the cell border as a result of reducing the ICI effect. This leads to an enhancement in throughput and an increase in the cell coverage area (capacity). FFR can be considered a combination of frequency reuse of 1 and 3 (see Figure 2.2a and b). Therefore, there is no need for frequency planning when FFR is used. However, the major disadvantage of using the FFR technique is the inefficient utilization of bandwidth (or resources) in the R3 zone (Figure 2.3), where only one-third of the bandwidth should be used.

2.5 FFR Deployment Classification

The main aim of FFR deployment is to address the ICI between cells. In terms of power, FFR implementation can be classified into hard and soft. This classification has been offered as a basis for discussion by [19] and was proposed by 3GGP LTE [20]. The configuration of hard FFR is the same FFR described in Section 2.4.2, whereby some subcarriers (or subchannels or part of the bandwidth) are not used to mitigate the ICI effect, which causes inefficient use of bandwidth. However, to increase the bandwidth utilization, soft FFR is proposed. In soft FFR, all the available subcarriers (or subchannels or bandwidth) are used, but those who cause the ICI will lower their transmission power, thus requiring power control entity. In addition, static and dynamic [21–23] techniques are proposed for hard and soft FFR deployment, as illustrated in Figure 2.4.

In static FFR, the system parameters are set earlier and remain fixed (unchanged) during a specific period of time; they are not changed as the cell traffic load changes. Conversely, in dynamic FFR the system parameters are dynamically changed frame by frame according to the cell load, which improves the overall system performance at the cost of increasing system complexity [24]. The adjustment of some FFR parameters affects system response, for example, the BS power level, FRF, zone size, and SINR or

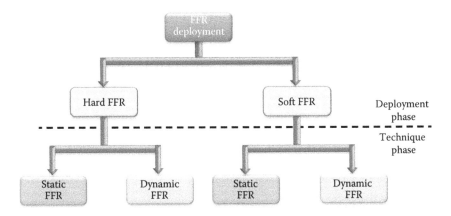

FIGURE 2.4
FFR deployment classification.

distance thresholds that allocate resources (subcarriers) to cell center users or cell border users [21,23]. The hard FFR is mostly categorized by researchers as traditional or conventional FFR [25–27], which could lead to confusion. In order to avoid confusion, in this book hard FFR is referred to as traditional FFR.

2.5.1 Dynamic FFR and Static FFR

The dynamic FFR technique is adapted in the OFDMA system, where the system data rate is increased when applying the dynamic subcarrier assignment (DSA) method [28]. This enhancement is due to the ability of the system to deal with each user channel independently, which increases the gain of user diversity. In contrast, the adaptive power allocation (APA) is another method used to enhance the system data rate [29]. APA shows how the subscribers in a given cell will share the limited power resources available at the base station. The APA controls the transmitted power by assigning different power levels to a set of subcarriers according to the measured channel state information (CSI). However, the integration of DSA and APA may result in greater system complexity, where the cross-layer design should be involved [30].

A cross-layer design was suggested by [31], where the problem of static allocation of subcarriers in traditional FFR has been considered. In the design, two methods are implemented to allocate subcarriers: radio network controller (RNC) and DSA. The cell area is partitioned into two virtual overlapping regions, which are called regular and super, and the subcarriers are also partitioned into two groups, called regular group and super group. The RNC allocates the subcarriers into all groups in the grid, while the DSA allocates the subcarriers to users in each cell. The results confirm that the dynamic subcarrier allocation increases system throughput compared with frequency reuse of 1, frequency reuse of 3, and traditional FFR. However, this enhancement is greatly noticeable in cells of small and medium radius up to 3 km. However, when the cell radius increases, the performance of the four schemes becomes equal. Similar to the former work, a dynamic FFR in the WiMAX system was applied by [32], where utility function was used instead of the DSA method to allocate subcarriers to users. In order to improve cell border throughput, the random access technique is used, where some of subcarriers are restricted to cell border users, which causes a reduction in total system throughput. Therefore, adaptive modulation and coding (AMC) is applied to increase the system throughput.

The DL subframe is partitioned into four zones [21], and the size of these zones is changed dynamically. The first three zones are restricted to serve users in the cell center, while the fourth zone is used to serve users in the cell border. Three types of modulation are defined in the cell center, where each zone can use only one type. The aim of the work is to increase the system throughput and decrease the outage probability in the cell edge, by changing the four zones' size frame by frame according to the system load. Two schemes were proposed to decide the size of the four zones: partial coordination and full coordination. In partial coordination, the size of the inner area zones is determined independently in each base station, while the fourth zone is determined centrally by the network core. However, in full coordination the size of all four zones is determined by the network core. The full coordination scheme shows more balance in resource management between the base stations than the partial coordination when the users are irregularly distributed in the network. Comparing the results of dynamic FFR and traditional FFR, the former reduces edge throughput outage by 0.45% and increases the total system throughput by 1.75 times when the users are irregularly distributed. However, the authors did not consider the unused segments in the traditional FFR DL subframe, which causes resource wastage.

A trade-off study between static and dynamic FFR is required [24]; basically, the dynamic FFR enhances system performance where the variation of load per cell is considered, but at the cost of increasing the overhead and complexity. In contrast, the static FFR controls the ICI with less overhead and complexity, but the static resources assignment leads to a decrease in throughput [24]. In the work of [33], the authors proposed a semidynamic ICI scheme to find common points between static and dynamic FFR. The proposed scheme increases system performance while maintaining ICI avoidance with less overhead. The new scheme deals with different cell loads, such as light, medium, and heavy loads. The obtained results indicate improved throughput of the cell edge users and minimize the blocking probability, especially when the users' distribution in the three sectors is unequal. However, the work addresses the ICI effect of the first tier (6 cells), while the second-tier ICI effect (12 cells) is ignored. In the heavy-load scenario, the subchannels are divided into six groups in each cell edge. These groups should be orthogonal to each other in the neighboring cells in order to reduce interference. This arrangement brings more system complexity when considering the second tier (12 cells) in the calculations.

2.6 FFR Resource Assignment Challenges in the OFDMA Network

Researchers tried many solutions to improve the performance of the FFR technique in the OFDMA system. However, most of the trends were concentrated on enhancing different issues, such as data rate, cell coverage area, cell capacity, resource utilization, and quality of service (QoS).

Resource allocation in OFDMA networks, such as WiMAX, is a complicated task where many issues are involved, such as ICI measurement, required resources or bandwidth, application type, and QoS requirements [34]. In addition, the resource assignment per zone or segment is also considered to be a problem by many researchers; for instance, the selection of the zone of operation and zone size (R1 and R3 zones) affects system performance. Furthermore, the WiMAX Forum did not specify the resource allocation structure for the FFR technique [16,18]. Most of the research focuses on solving the problem of resource assignment in the OFDMA system through two levels: the local level, which involves allocating the resources to subscribers at the base station level, and the global level, which involves allocating the resources to all base stations in the grid at the network level. In some implementations, these two levels cooperate together to increase system performance. In this section, some of the proposed solutions are presented based on the techniques used.

2.6.1 Resource Allocation Based on Distance

This approach has been adapted by [35–37], where the cell center area and the cell border area are specified according to a predetermined distance from the base station. Users near the base station can be served by the cell center area (i.e., R1 zone), while users far away from the base station can be served by the cell border area (i.e., R3 zone). The channel estimation is based on the distance of the users from the base station rather than considering noise and interference signals from the other base stations in the network.

A trade-off study between interference avoidance and interference averaging was presented by [36], which results in optimal frequency planning. The frequency reuses of 1 and 3, as well as the traditional FFR technique, have been implemented using PUSC with two different modes: the same permutation base and a different permutation base. In the study, different ratios of the cell center area were examined to find the optimal ratio (i.e., optimal inner area radius) that gives the highest data rate. At this ratio, the performance of the traditional FFR is compared with that of the other two models in terms of outage probability and network throughput. The results of the study show that the outage probability of frequency reuse of 1 becomes unfeasible for both permutation modes, whereas only the same permutation mode is feasible for a frequency reuse of 3. Meanwhile, the highest system throughput is achieved by the traditional FFR technique. However, the authors in [37] show that traditional FFR performs well in low-population-density environments, such as rural areas. The average number of bits per symbol per subcarrier was used as a metric to analyze the traditional FFR technique in rural areas. The inner area radius was determined to guarantee better traditional FFR performance. It shows that the inner cell radius affects the number of bits that can be transmitted per subcarrier as a result of using different modulation types. The results were compared with those of classical frequency reuse planning, where the cell capacity and the number of bits per subcarrier are increased when considering the traditional FFR technique. However, the authors did not address the resource wastage in the R3 zone due to the two unused segments in the traditional FFR technique.

2.6.2 Resource Allocation Based on SINR

Resource assignment in this approach considers the interference from other cells, whereby system parameters such as zone size are affected by the measured SINR. The ratio of the cell center area and the ratio of the cell border area are specified based on a predetermined SINR threshold. Users near the base station that enjoy high SINR can be served in the cell center area (i.e., R1 zone), whereas users far away from the base station with low SINR can be served in the cell border area (i.e., R3 zone). This method is useful when analyzing target cell performance while taking into consideration the interference effect of the adjacent cells.

The problem of resource assignment was presented by [38,39], where the SINR as a metric is used to decide which zone will assign resources to a particular user. The studies compare three schemes; frequency reuse of 1, frequency reuse of 3 (sectoring), and the traditional FFR technique. In [39], the highest throughput was achieved by frequency reuse of 1 compared with traditional FFR and reuse 3. Nevertheless, the coverage area in the traditional FFR was enhanced better than that of reuse 1, while it achieved almost the same coverage area of reuse 3. However, in [38] the zone size becomes critical in enhancing the cell coverage area where different R3 zone sizes were examined. The obtained results show that the traditional FFR improves both coverage area and throughput, while it achieves a coverage area almost similar to that of the reuse 3 scheme. Conversely, the resource wastage in the traditional FFR DL subframe due to the unused two segments in the R3 zone was not considered in these studies.

The FFR concept is used in the OFDMA system as proposed by [40]. The data subcarriers are divided into two groups: a super group to serve users near the base station and a regular group to serve users at the cell border. The regular group subcarriers are equally and orthogonally divided into three sectors. The two groups are separated in time and are not used together. Therefore, the intracell interference is omitted and the ICI is mitigated; hence, the co-channel effect is reduced. In this work, two approaches are used, distance base and SINR base, whereby the system is switched from one group to another based on two threshold criteria: SINR and distance. The results of the two approaches were compared with frequency reuse of 1, and the system throughput was increased when using the SINR threshold. However, the design was limited to offering regular group subcarriers to six users in the cell border. If the number of users in the cell border is increased, then there are insufficient subcarriers to serve them.

The SINR approach considers the interference from other base stations when assigning resources. However, in some cases users near the cell border may have enough SINR value, which forces the base station to consider these users as a member of the cell center area. In order to solve this issue, a combination of two threshold types, namely, distance and SINR, were presented by [41]. The purpose of this combination is to increase the performance of cell border users, and thus to ensure that users near the cell border in some frames with good channel conditions will not be served by the cell center area (R1 zone) in the next frame. The results proved that the new combination scheme improves the outage probability of the system compared with the results of distance and SINR approaches individually. However, the resource wastage in the traditional FFR due to two unused segments in the R3 zone was not considered in the optimization.

Applying the concept of FFR is not limited to 3G and 4G; it has also been adapted in access point (AP) deployment. The ergodic capacity method was proposed by [42], to assign users to either the R1 or R3 zone. The target of the work is to determine when or where to switch from the R1 zone to the R3 zone, that is, to make a fast and accurate decision to assign users to either the R1 or R3 zone based on numerical analysis (the ergodic capacity method). In [42], the performance of FFR AP was examined under different shadowing values and path loss effect, unlike other works where the shadowing is assumed equal to a constant value [43]. The ergodic capacity method is used as performance measurement to find an optimal strategy to assign users to either the R1 zone or the R3 zone, which results in an increase in system capacity. However, the unused segments in the DL subframe of the traditional FFR technique were not considered in this work, which yields resource wastage.

An analytic approach based on the fluid model was proposed for code-division multiple access (CDMA) networks [44,45] and adapted by [46] to evaluate the performance of the OFDMA cellular network. In this approach, three models were simulated using the Monte Carlo simulation: frequency reuse of 1, traditional FFR, and two-level power control (TLPC). Through the analytic approach, it was proved that the fluid model can be used to simulate the aforementioned three models in terms of SINR, system capacity, and throughput. However, the implementation of traditional FFR in the fluid analytic approach emphasized exploiting the existing resource in the R1 and R3 zone, whereas the unused resources in the R3 zone were not considered.

In [47], the signal-to-noise ratio (SNR) is used instead of using SINR to allocate resources, unlike traditional FFR, where two segments out of three in the R3 zone are used in the cell border to increase the spectral efficiency of the system. The utilization of the two segments simultaneously in this work leads to interference between frequencies in the cell border. In order to benefit the interfered segment, the interference suppression strategy is proposed, by assigning different modulation types for the interfered subcarriers of the neighboring cells (i.e., nonuniform data rate). It shows that when the interfered signal increases, the required SNR level for a particular modulation type is reduced. This results in minimizing the transmission power of the base stations. However, the exploitation of another segment in the R3 zone serves users in the cell border, which increases the ICI, especially when the number of users in the cell border is increased. Besides, the third segment in R3 was not considered in the proposed strategy, which results in wastage in resources.

2.6.3 Resource Allocation Based on Optimization Analysis

Resource assignment issues in the FFR technique can be solved by the optimization approach. The system parameters, such as zone size, FRF, and the area size of the cell center and cell border, are computed mathematically as optimization problems. Optimization analysis helps us to find the perfect values of the system parameters while increasing the complexity of the system [48].

In [40,49], the distance and SINR thresholds are calculated based on finding the optimal switching point between the R1 and R3 zones, which maximizes the SINR and minimizes co-channel interference (CCI) at the cell border. The optimal distance threshold was found by [50], in terms of two types of scheduler, namely, round-robin (RR) and maximum SINR (MSINR). Throughput in the proposed FFR with an optimal distance threshold outperformed the traditional FFR (fixed threshold). In addition, the results demonstrate that when the number of users increases, there will be an increase in throughput, while the optimal distance threshold decreases (cell center radius). However, the authors did not consider the unused segments in the DL subframe of the traditional FFR technique, which results in wastage of resources.

The problem of resource assignment in FFR was introduced by [48] as an optimization problem, to obtain the optimal trade-off between the cell edge FRF and ICI avoidance. In the proposed optimization, a different set of subcarriers are assigned to users in neighboring cells in order to avoid ICI.

However, the proposed optimization failed to assign subcarriers to two users sharing the same good channel quality. Moreover, it brings high computational complexity, which makes it unsuitable for implementation. Therefore, a classical optimization problem is proposed as an alternative solution to reducing complexity. The finding of the optimal FRF resulted in an increase of throughput of the cell border users at a cell center radius equal to two-thirds the cell radius. The authors of [51] proposed an optimal solution to increase the throughput and spectral efficiency for cell border users. The proposed solution divides the optimization problem into two subproblems to avoid complexity. In this work, three FRFs are used to reduce the ICI effect: 1, 2/3, and 1/3. The implementation of these three FRFs requires partitioning the DL subframe into three zones with a suitable number of segments, which results in unused resources (slots). Thus, the proposed enhancement in this work did not consider resource wastage in the DL subframe as a result of partitioning the DL subframe.

The optimization problems were repeated in a different manner by [52], where the optimization problem of power allocation and subcarrier assignment in traditional FFR was addressed by proposing a new algorithm. The proposed algorithm was applied in two cells, and the subcarriers in each cell were divided into two parts. The first part was used by cell center users, while the second part was subdivided into two groups. Only one group is used by cell border users in each cell; thus, the ICI of the neighbor cell border is avoided. However, high computational complexity was observed when applying the proposed algorithm since it required a large number of operations. Therefore, in [53] the complexity is averted by proposing a suboptimal resource allocation algorithm. As a result, the system spectral efficiency is improved. However, in the previously mentioned algorithms, part of the subcarriers, called forbidden subcarriers, are not used in the DL subframe, which yields resource wastage.

2.6.4 Resource Allocation Based on Load Balance

In the previous resource assignment solutions, the selection of system parameters depends on specific metrics, such as distance threshold, SINR threshold, or an optimization threshold, without considering the available resources (slots) in the intended frame part (zone). The previous thresholds are strict in specifying the zone of operation for a particular user; the user will be assigned to one of the zones (the R1 or R3 zone) according to the specific threshold value. If the available free slots in the target zone are considered when making a decision to determine the zone of operation, then that will increase resource utilization and decrease probability of user rejection.

The load balance model suggested by [54] aims to assign resources based on two metrics: channel condition and the available free slots in the target zone. Three models are implemented in this work, namely, distance based, SINR based, and load balance based. In addition, and in order to evaluate the performance of these three models, different FRFs are examined, which requires changing the zones' size. The resource allocation in the first two models (SINR and distance) performed poorly compared with the load balance model, where the former two models could not efficiently assign resources for various zone sizes. The results show that the load balance model overcomes the SINR and distance models in terms of blocking probability and achievable bit rate; hence, the system tries to utilize the available resources efficiently. Similarly, an improvement was shown by [55], where the three former models are implemented (distance, SINR, and load balance). The channel status is measured based on the ray-tracing propagation model, which makes the channel estimation close to reality. The load balance improves the performance of WiMAX networks in terms of blocking ratio and offered bit rate. It is possible to conclude that when the zones' sizes are dynamically changed, it is better to consider the load balance approach as a metric to assign resources. However, the previous works did not consider resource wastage in the traditional FFR as a result of unused segments in the R3 zone.

2.6.5 Resource Allocation Based on Quality of Service

Another method for resource allocation in the OFDMA network is based on QoS. The zone assignment algorithm proposed by [56] is based on zone time partitioning with less complexity in order to maximize resource utilization in the WiMAX DL subframe. The service flow takes into account the relative SINR and zone size when assigning resources to a user. However, the zone assignment algorithm did not consider the unused segments in the traditional FFR technique, which causes a reduction in resource utilization.

A client-centric FFR (CC-FFR) algorithm was proposed by [57] in the OFDMA network to solve the problem of resource management in the traditional FFR. Instead of using base station coordination to manage resources, a channel quality indicator (CQI) is used as a metric to assign resources at the base station level. The CC-FFR measures the ICI level at the cell border users, and based on this measurement, a suitable subchannel with low interference is allocated to the cell border users. As a result, the CC-FFR increases the spectral efficiency of the system compared with traditional FFR. However, the proposed CC-FFR did not consider resource wastage in the traditional FFR DL subframe, which occurs as a result of unused segments in the R3 zone.

In [58], the authors proposed a resource allocation scheme in the multi-cell WiMAX system by implementing a radio resource controller (RRC) and radio resource agent (RRA) using the FFR technique. The RRC was used to control the ICI while maintaining the service flow types and their QoS requirements, while the RRA was used to allocate the slots in each cell fairly between cell center users and the cell border. The proposed work includes the real-time and non-real-time service flow applications. The results show that the overall system throughput is increased while maintaining the QoS required. However, even though the results enhance traditional FFR performance, the unused segments in the R3 zone were not considered in the enhancement.

2.6.6 Resource Allocation Based on Frame Partitioning

OFDMA allows for dividing of the DL subframe into several zones, since the resources are defined by two dimensions, time and frequency. Each zone may use a different permutation mode and may operate with part of the available bandwidth. In the time-division duplexing (TDD) WiMAX frame structure [13], the available subcarriers are divided into groups of subchannels. If the PUSC permutation mode is used, then six groups of subchannels can be formed, where every two groups form a segment. In frame partitioning, the available frequency bandwidth is divided into several subchannels (or subbands), and these subchannels can be used by a zone or a segment to avoid ICI.

A novel frequency partitioning approach has been proposed by [59], whereby the DL subframe is divided into four segments that correspond to four frequency bands, F1, F2, F3, and F4, and each band contains an equal number of subchannels. The first band (F1) is used in the cell center and cannot to be used in the borders of the cells in the grid. Another three bands (F2, F3, and F4) are used in the cell border of the three adjacent cells. However, each of these three bands can be used in the cell center in order to increase cell center capacity. In this arrangement, the ICI in the cell borders is reduced, as well as in the cell center, because only one frequency band is repeated (i.e., F1). Furthermore, a new subcarrier allocation mechanism is proposed in this approach, where the users in the cell center can use the cell border subcarriers as long as there are enough places to serve new users. The results were compared with traditional FFR and without FFR (i.e., frequency reuse of 1). As a result, system throughput is enhanced due to an increase in the SINR value, especially for users at the cell border. Although this new frequency partitioning approach reduces the ICI and enhances the SINR, technically the DL subframe should be divided into four segments where only two of them can be used in each cell. Thus, in each DL subframe of each base station there are two unused segments. These cause a wastage of resources, which limits the number of served users.

The DL subframe may be divided into four zones to control the ICI effect as suggested by [60], where the cluster size equals four cells, and the total bandwidth is divided into four frequency bands to form four FRFs called 1, 1/2, 1/3, and 1/4. The four cells can reuse three types of frequency bands (1, 1/2, 1/3), while the fourth band (1/4) is used in each cell border, thus minimizing the ICI effect. Two different methods termed radio distance (RD) and bandwidth efficiency (BWE) are presented to assign resources to users. These two methods are used to determine the zone of operation, and they increase the throughput of the system compared with static zone assignment (traditional FFR). Specifically, the BWE performs better than the RD because in BWE the zone selection decision is made based on maximizing the bandwidth utilization, whereas in RD the zone selection decision is made based on the received signal strength indicator (RSSI). However, three of the frequency bands are allowed to be reused in the neighboring cells, which means that the impact of ICI is still there. In addition, the partitioning of the DL sub-frame into four zones increases the unused resources in the DL subframe.

Frequency planning design in OFDMA cellular networks is difficult to achieve, especially in a specific geographical area and with the increasing demand for various E-services. Network designers may try different strategies in order to arrive at an optimal solution by reducing the co-channel effect and increasing spectral efficiency. A multiantenna terminal approach termed intercell spatial demultiplexing (ISD) was proposed by [61] to reduce the impact of CCI in the FFR technique, as well as in the frequency reuse of 1 and 3. The multiantenna performs as a spatial demultiplexing, which distinguishes between both the interfering signals and the desired signal. The ISD shows much better spectral efficiency (particularly in the cell border) than the maximal ratio combining (MRC) method. In multiantenna terminals, MRC is a signal detection method where the signal from each channel is added together. However, the design of FFR was created with two types of FRFs: 2/3 and 1/3. This design requires the DL subframe to be partitioned into a suitable number of segments to match the required FRFs, which leads to resource wastage in the DL subframe.

2.7 Resource Enhancement in the FFR Technique

In Section 2.6, the performance enhancement of the FFR technique was discussed through six types of solutions. These solutions included modifying the system parameters, such as threshold selection, zone size selection, power control, quality of service requirement, frame partitioning, and dynamic resource allocation. However, each of these solutions attempted to satisfy the most important factors in deploying FFR base stations, such as bandwidth utilization and resource utilization, while controlling the ICI as much as possible.

Some solutions focus on finding the optimal threshold, such as distance and SINR, to increase the data rate and reduce users' rejection. SINR as a metric is better than distance since it considers the interference from other base stations in the network when assigning resources. Other solutions are concerned about finding the optimal system parameters to increase the base station performance, such as selection of the zone of operation and zone size. Some researchers resort to sacrificing bandwidth by using the frame partition method, where looser FRF is applied to mitigate the effect of interference. Using looser FRF reduces the interference but reduces the spectral efficiency as well. On the other hand, using tighter FRF increases the spectral efficiency but increases the interference between neighboring base stations, especially at the cell border.

Dynamic FFR is considered by many researchers to increase the performance of OFDMA networks, where the system parameters, like zone size, are changed dynamically according to the cell traffic load. However, dynamically changing zone sizes requires more control signals, which results in more system complexity; besides, the zone sizes in every base station in the network must be uniform to avoid interzone interference, which leads to more complexity.

Radio resource utilization in the OFDMA network is always the focus of researchers. Solutions to increase the radio resource utilization are required to satisfy the growth number of broadband wireless users. In general, a good solution should be with minimum changes to reduce the complexity and cost, such as avoiding the following: changing the FFR network infrastructure, using more devices or power control entity, increasing the interference between the base stations, increasing the complexity (dynamic FFR), and losing more bandwidth, such as in frame portioning method.

2.8 Summary

The main issues of FFR base station deployment in OFDMA cellular networks have been reviewed in this chapter. First, the problem of ICI in cellular networks is explained. A range of available solutions are then presented to solve the interference problem, and among these solutions the FFR technique is used. FFR can be classified into two types: hard (traditional FFR) and soft FFRs. The advantages and disadvantages of these two types are discussed, and the problems of FFR base station deployment are illustrated. The consequences are presented in terms of the solutions used in improving the performance of the traditional FFR technique by discussing the work done in that area.

Most of the solutions focused on increasing the system performance by increasing radio resource utilization and keeping the interference at an acceptable level. This fact leads to either not using part of the bandwidth or using intelligent dynamic algorithms to control the interference. However, the former leads to more wastage in radio resources, and the latter increases the system complexity. An important goal of the FFR technique is to avoid interference between adjacent cells, where parts of the subchannels (bandwidth) are not used. These subchannels are occupied by segments B and C in the traditional FFR technique (Figure 2.3). This is mandatory to ensure that different bands of frequency (subchannels) are used in each cell border to mitigate the ICI effect. However, the enhancements in the previous works did not consider the unused segments (B and C) in traditional FFR, which causes resource wastage.

References

1. D. Pareit, B. Lannoo, I. Moerman, and P. Demeester, The history of WiMAX: A complete survey of the evolution in certification and standardization for IEEE 802.16 and WiMAX, *IEEE Communications Surveys and Tutorials*, vol. 14, pp. 1183–1211, 2012.
2. IEEE-Sta, IEEE standard for local and metropolitan area networks part 16: Air interface for fixed broadband wireless access systems, in IEEE Standard 802.16-2001, 2002, pp. 0_1-322.
3. IEEE-Sta, IEEE standard for local and metropolitan area networks part 16: Air interface for fixed broadband wireless access systems, in IEEE Standard 802.16-2004 (revision of IEEE Standard 802.16-2001), 2004, pp. 0_1-857.
4. IEEE-Sta, IEEE standard for local and metropolitan area networks part 16: Air interface for fixed and mobile broadband wireless access systems amendment 2: Physical and medium access control layers for combined fixed and mobile operation in licensed bands and corrigendum 1, in IEEE Standard 802.16e-2005 and IEEE Standard 802.16-2004/Cor 1-2005 (amendment and corrigendum to IEEE Standard 802.16-2004), 2006, pp. 0_1-822.
5. V. MacDonald, The cellular concept, *Bell System Technical Journal*, vol. 58, pp. 15–41, 1979.
6. K. R. Manoj, *Coverage Estimation for Mobile Cellular Networks from Signal Strength Measurements*, University of Texas, Dallas, 1999.
7. J. G. Andrews, A. Ghosh, and R. Muhamed, *Fundamentals of WiMAX: Understanding Broadband Wireless Networking*, Pearson Education, London, 2007.
8. R. Ramanathan and R. Rosales-Hain, Topology control of multihop wireless networks using transmit power adjustment, in *INFOCOM 2000: Proceedings of the Nineteenth Annual Joint Conference of the IEEE Computer and Communications Societies*, Tel Aviv, Israel, 2000, vol. 2, pp. 404–413.
9. S. Glisic and B. Lorenzo, *Advanced Wireless Networks: Cognitive, Cooperative and Opportunistic 4G Technology*, Wiley, Hoboken, NJ, 2009.
10. A. Goldsmith, *Wireless Communications*, Cambridge University Press, Cambridge, 2005.
11. L. Korowajczuk, *LTE, WIMAX and WLAN Network Design, Optimization and Performance Analysis*, Wiley, Hoboken, NJ, 2011.
12. M. C. Necker, Scheduling constraints and interference graph properties for graph-based interference coordination in cellular OFDMA networks, *Mobile Networks and Applications*, vol. 14, pp. 539–550, 2009.
13. Y. Zhang and H.-H. Chen, *Mobile WiMAX: Toward Broadband Wireless Metropolitan Area Networks*, CRC Press, Boca Raton, FL, 2007.

14. M. Maqbool, M. Coupechoux, and P. Godlewski, Comparative study of reuse patterns for WiMAX cellular networks, Technical report, TELECOM ParisTech, 2008.
15. K. Begain, G. I. Rozsa, A. Pfening, and M. Telek, Performance analysis of GSM networks with intelligent underlay-overlay, in *Proceedings of the ISCC 2002 Seventh International Symposium on Computers and Communications*, Japan, 2002, pp. 135–141.
16. Forum, Mobile WiMAX—Part I: A technical overview and performance evaluation, 2006, p. 53.
17. 3GPP-R1-050841, *Further Analysis of Soft Frequency Reuse Scheme*, HuaWei, London, 2005.
18. Forum, Mobile WiMAX—Part II: A comparative analysis, 2006, p. 47.
19. IEEE-C802.16m-08/017, Frame structure to support inter-cell interference mitigation for downlink traffic channel using Co-MIMO and FFR, IEEE 802.16 Broadband Wireless Access Working Group, 2008.
20. 3GGP-R1-050764, Inter-cell interference handling for E-UTRA, Ericsson, London, 2005.
21. S. Hamouda, C. Yeh, J. Kim, S. Wooram, and D. S. Kwon, Dynamic hard fractional frequency reuse for mobile WiMAX, in *IEEE International Conference on Pervasive Computing and Communications, PerCom 2009*, Galveston, Texas, 2009, pp. 1–6.
22. 3GGP-R1-050896, Description and simulations of interference management technique for OFDMA based E-UTRA downlink evaluation, QUALCOMM Europe, 2005.
23. 3GPP-R1-060545, Some clarifications on soft frequency reuse scheme, Huawei, Denver, 2006.
24. S.-E. Elayoubi, O. Ben Haddada, and B. Fourestie, Performance evaluation of frequency planning schemes in OFDMA-based networks, *IEEE Transactions on Wireless Communications*, vol. 7, pp. 1623–1633, 2008.
25. C.-N. Lee, Y.-T. Chen, Y.-C. Kao, H.-H. Kao, and S. Haga, Layered video multicast using fractional frequency reuse over wireless relay networks, *EURASIP Journal on Wireless Communications and Networking*, vol. 2012, pp. 1–16, 2012.
26. S.-W. Kim and Y.-H. Lee, Adaptive MIMO mode and fast cell selection with interference avoidance in multi-cell environments, in *Fifth International Conference on Wireless and Mobile Communications, ICWMC'09*, Cannes, France, 2009, pp. 163–167.
27. X. Li, H. Jin, J. Jiang, S. Hou, M. Peng, and G. Wang, A gradient projection based self-optimizing algorithm for inter-cell interference coordination in downlink of DMA networks, in *7th International ICST Conference on Communications and Networking in China, ChinaCom 2012*, Kunming, China, 2012.
28. G. Song and Y. Li, Cross-layer optimization for OFDM wireless networks—Part I: Theoretical framework, *IEEE Transactions on Wireless Communications*, vol. 4, pp. 614–624, 2005.
29. R. Bo, Q. Yi, and C. Hsiao-Hwa, Adaptive power allocation and call admission control in multiservice WiMAX access networks [Radio Resource Management and Protocol Engineering for IEEE 802.16], *IEEE Wireless Communications*, vol. 14, pp. 14–19, 2007.

30. B. Rong, Y. Qian, and K. Lu, Integrated downlink resource management for multiservice WiMAX networks, *IEEE Transactions on Mobile Computing*, vol. 6, pp. 621–632, 2007.
31. S. H. Ali and V. C. Leung, Dynamic frequency allocation in fractional frequency reused OFDMA networks, *IEEE Transactions on Wireless Communications*, vol. 8, pp. 4286–4295, 2009.
32. Z. Mohades, V. T. Vakili, S. M. Razavizadeh, and D. Abbasi-Moghadam, Dynamic fractional frequency reuse (DFFR) with AMC and random access in WiMAX system, *Wireless Personal Communications*, vol. 68, pp. 1871–1881, 2013.
33. L. Li, D. Liang, and W. Wang, A novel semi-dynamic inter-cell interference coordination scheme based on user grouping, in *IEEE 70th Vehicular Technology Conference, VTC Fall 2009*, Anchorage, AK, 2009, pp. 1–5.
34. L. Nuaymi, *WiMAX: Technology for Broadband Wireless Access*, Wiley, Hoboken, NJ, 2007.
35. C. Tian, J. Jin, and X. Zhang, Evaluation of mobile WiMAX system performance, in *IEEE 68th Vehicular Technology Conference, VTC Fall 2008*, Calgary, Canada, 2008, pp. 1–5.
36. H. Jia, Z. Zhang, G. Yu, P. Cheng, and S. Li, On the performance of IEEE 802.16 OFDMA system under different frequency reuse and subcarrier permutation patterns, in *IEEE International Conference on Communications, ICC'07*, Glasgow, Scotland, 2007, pp. 5720–5725.
37. R. Giuliano, C. Monti, and P. Loreti, WiMAX fractional frequency reuse for rural environments, *IEEE Wireless Communications*, vol. 15, pp. 60–65, 2008.
38. Y. Zhou and N. Zein, Simulation study of fractional frequency reuse for mobile WiMAX, in *IEEE Vehicular Technology Conference, VTC Spring 2008*, Singapore, 2008, pp. 2592–2595.
39. V. L. S. V. M. Hunukumbure and V. S. Vadgama, Simulation study of fractional frequency reuse in WiMAX networks, *Fujitsu Scientific and Technical Journal*, vol. 44, pp. 318–324, 2008.
40. H. Lei, L. Zhang, X. Zhang, and D. Yang, A novel multi-cell OFDMA system structure using fractional frequency reuse, in *IEEE 18th International Symposium on Personal, Indoor and Mobile Radio Communications, PIMRC 2007*, Athens, Greece, 2007, pp. 1–5.
41. Y. Chen, W. Wang, T. Li, X. Zhang, and M. Peng, Fractional frequency reuse in mobile WiMAX, in *Third International Conference on Communications and Networking in China, ChinaCom 2008*, Hangzhou, China, 2008, pp. 276–280.
42. B. Pijcke, M. Gazalet, M. Zwingelstein-Colin, and F.-X. Coudoux, An accurate performance analysis of an FFR scheme in the downlink of cellular systems under large-shadowing effect, *EURASIP Journal on Wireless Communications and Networking*, vol. 2013, pp. 1–14, 2013.
43. C. Sankaran, F. Wang, and A. Ghosh, Performance of frequency selective scheduling and fractional frequency reuse schemes for WiMAX, in *IEEE 69th Vehicular Technology Conference, VTC Spring 2009*, Barcelona, Spain, 2009, pp. 1–5.
44. J.-M. Kelif and E. Alman, Downlink fluid model of CDMA networks, in *IEEE 61st Vehicular Technology Conference, VTC Spring 2005*, Stockholm, Sweden, 2005, pp. 2264–2268.

45. J.-M. Kelif, M. Coupechoux, and P. Godlewski, Spatial outage probability for cellular networks, in *IEEE Global Telecommunications Conference, GLOBECOM'07*, New Orleans, LA, 2007, pp. 4445–4450.

46. P. Godlewski, M. Maqbool, M. Coupechoux, and J.-M. Kélif, Analytical evaluation of various frequency reuse schemes in cellular OFDMA networks, in *Proceedings of the 3rd International Conference on Performance Evaluation Methodologies and Tools*, Greece, 2008, p. 32.

47. R. Ghaffar and R. Knopp, Fractional frequency reuse and interference suppression for OFDMA networks, in *Proceedings of the 8th International Symposium on Modeling and Optimization in Mobile, Ad Hoc and Wireless Networks (WiOpt)*, France, 2010, pp. 273–277.

48. M. Assaad, Optimal fractional frequency reuse (FFR) in multicellular OFDMA system, in *IEEE 68th Vehicular Technology Conference, VTC Fall 2008*, Calgary, Canada, 2008, pp. 1–5.

49. A. Najjar, N. Hamdi, and A. Bouallegue, Efficient frequency reuse scheme for multi-cell OFDMA systems, in *IEEE Symposium on Computers and Communications, ISCC 2009*, Alexandria, Egypt, 2009, pp. 261–265.

50. Z. Xu, G. Y. Li, C. Yang, and X. Zhu, Throughput and optimal threshold for FFR schemes in OFDMA cellular networks, *IEEE Transactions on Wireless Communications*, vol. 11, pp. 2776–2785, 2012.

51. L. Fang and X. Zhang, Optimal fractional frequency reuse in OFDMA based wireless networks, in *4th International Conference on Wireless Communications, Networking and Mobile Computing, WiCOM'08*, Dalian, China, 2008, pp. 1–4.

52. N. Ksairi, P. Bianchi, P. Ciblat, and W. Hachem, Resource allocation for downlink cellular OFDMA systems—Part I: Optimal allocation, *IEEE Transactions on Signal Processing*, vol. 58, pp. 720–734, 2010.

53. N. Ksairi, P. Bianchi, P. Ciblat, and W. Hachem, Resource allocation for downlink cellular OFDMA systems—Part II: Practical algorithms and optimal reuse factor, *IEEE Transactions on Signal Processing*, vol. 58, pp. 735–749, 2010.

54. I. N. Stiakogiannakis and D. I. Kaklamani, Fractional frequency reuse techniques for multi-cellular WiMAX networks, in *IEEE 21st International Symposium on Personal Indoor and Mobile Radio Communications (PIMRC)*, Turkey, 2010, pp. 2432–2437.

55. I. N. Stiakogiannakis, G. E. Athanasiadou, G. V. Tsoulos, and D. I. Kaklamani, Performance analysis of fractional frequency reuse for multi-cell WiMAX networks based on site-specific propagation modeling [wireless corner], *IEEE Antennas and Propagation Magazine*, vol. 54, pp. 214–226, 2012.

56. M. Einhaus, A. Mäder, and X. Pérez-Costa, A zone assignment algorithm for fractional frequency reuse in mobile WiMAX networks, in *NETWORKING 2010*, Springer, Berlin, 2010, pp. 174–185.

57. S. Geirhofer and O. Oyman, Client-centric fractional frequency reuse based on user cooperation in OFDMA networks, in *42nd Annual Conference on Information Sciences and Systems, CISS 2008*, Princeton, NJ, 2008, pp. 95–100.

58. T. Ali-Yahiya and H. Chaouchi, Fractional frequency reuse for hierarchical resource allocation in mobile WiMAX networks, *EURASIP Journal on Wireless Communications and Networking*, vol. 2010, p. 7, 2010.

59. S. S. Han, P. Jongho, L. Tae-Jin, and H. G. Ahn, A new frequency partitioning and allocation of subcarriers for fractional frequency reuse in mobile communication systems, *IEICE Transactions on Communications*, vol. 91, pp. 2748–2751, 2008.
60. J. J. J. Roy and V. Vaidehi, Analysis of frequency reuse and throughput enhancement in WiMAX systems, *Wireless Personal Communications*, vol. 61, pp. 1–17, 2011.
61. J. Chang, J. Heo, and W. Sung, Cooperative interference mitigation using fractional frequency reuse and intercell spatial demultiplexing, *Journal of Communications and Networks*, vol. 10, pp. 127–136, 2008.

3

---

*Fractional Frequency Reuse Design
and Techniques for WiMAX*

3.1 Introduction

In wireless telecommunication, there is still work required to improve connection quality and capacity. Cellular networks increase capacity by frequency reuse. On the other hand, connection quality is a matter of enhancing
users' signal strength. In order to combine connection quality and capacity,
an effective base station is needed. WiMAX technology has emerged as a
new broadband wireless communication system with high capabilities,
which enable it to become a strong alternative to modern communication
systems [1]. WiMAX with mobile version (802.16e) has distinct capabilities
through the outstanding design of its base stations.

WiMAX base stations are flexible in design and offer scalable bandwidth
options (from 1.25 to 20 MHz), different subchannelizing options, and duplexing options (time- and frequency-division duplex) [2]. Moreover, WiMAX
supports fixed, nomadic, and mobile users [3]. In mobile applications, the
number of subcarriers (or fast Fourier transform [FFT] sizes) can vary from
128 to 2048 subcarriers [2]. The WiMAX physical layer uses orthogonal frequency-division multiplexing (OFDM) as a basic technique for data transmission and orthogonal frequency-division multiple access (OFDMA) as
resource allocation method, which makes it a strong candidate for network
deployment [4]. In addition, the flexibility of WiMAX base stations enables
the use of different frequency reuse factors, which results in different network deployment schemes, as is the case in cellular networks.

This chapter illustrates the basic technical issues of WiMAX base stations
related to the contributions of this work. The rest of the chapter is organized
as follows: Section 3.2 introduces the concept of OFDM and OFDMA. Section
3.3 explains the frame structure of WiMAX. The design parameters of different subcarrier permutation methods are discussed in Section 3.4. Further
discussion on WiMAX frame partition is presented in Section 3.5 through
the concept of zoning. WiMAX supports various types of modulation and
coding; these types are presented in Section 3.6. The quality of service (QoS)

classes in WiMAX and the concept of the media access control (MAC) scheduler are explained in Section 3.7. A good choice to deploy WiMAX base stations in cellular networks is based on the fractional frequency reuse (FFR) technique. The concept and the design parameters of this technique are discussed in Section 3.8. Finally, Section 3.9 summarizes this chapter.

3.2 OFDM and OFDMA in WiMAX

WiMAX technology supports line-of-sight (LOS) and non-line-of-sight (NLOS) conditions using the OFDM technique. OFDM divides the high-data-rate stream into several parallel lower-data-rate streams; modulates each stream on a separate carrier, often called a subcarrier; and transmits all the subcarriers simultaneously in parallel form [5]. The standard defined three types of subcarriers in the OFDM symbol are shown in Figure 3.1 [3]; these are data subcarriers, pilot subcarriers, and null subcarriers. There are two types of null subcarriers: direct current (DC) and guard subcarriers. The frequency of the DC subcarrier is equal to the RF center frequency of the transmitting base station and is not modulated (null). The DC subcarrier is used to simplify digital-to-analogue and analogue-to-digital converter operations. In contrast, the guard subcarriers are not used for transmission; instead, they are used as frequency guard bands. The pilot subcarriers are used for channel estimation and tracking. However, data subcarriers are used to carry user data.

Data subcarriers are not overlapped during each OFDM symbol time; they are selected in a certain way to ensure their orthogonality over the symbol duration time, such that each subcarrier has a specific frequency bandwidth to make sure that there is fixed space between subcarriers [6]. The orthogonality of subcarriers eliminates intercarrier interference and manages to isolate sets of subcarriers in a group of subchannels. These subchannels are orthogonal to each other; in fact, this is the concept of the OFDMA method. OFDM is classified as a type of modulation technique, whereas OFDMA is a channel access method [7]. The WiMAX base OFDMA physical layer is a

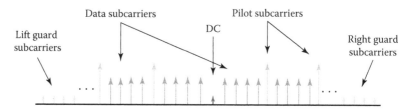

FIGURE 3.1
OFDM subcarrier types in WiMAX.

multiple access technique, whereby the subcarriers are divided into several logical subchannels, and each subchannel can be assigned to user data. The subchannelizing is allowed in both link directions, that is, uplink (UL) and downlink (DL) [8], while in the WiMAX base OFDM physical layer it is only allowed in the UL direction. OFDMA is flexible in managing the resources in the WiMAX frame, whereby different users can be allocated to different subsets of subcarriers, which enables WiMAX to improve system capacity by exploiting frequency diversity and multiuser diversity [9].

IEEE 802.16–2004 and 802.16e–2005 refer to fixed and mobile WiMAX technology, respectively. Table 3.1 shows the specifications of the fixed and mobile WiMAX standards for different channel bandwidths [3].

The MAC protocol architecture of IEEE 802.16–2004 and IEEE 802.16e is designed to support two types of network topologies (sometimes called modes): mesh and point-to-multipoint (PMP) [3]. In PMP topology, the client stations (users) are connected with each other through the base station, and no direct connections between the clients are allowed. The connection from the base station to the client station is denoted as the DL connection, while the connection from the client to the base station is denoted as the UL connection [10]. On the other hand, clients can communicate with each other with or without invoking the base station in mesh topology; therefore, there is no concept of DL or UL in mesh topology [11].

TABLE 3.1

Fixed and Mobile WiMAX System Parameters

Parameter	Fixed WiMAX	Mobile WiMAX			
Channel bandwidth (MHz)	3.5	1.25	5	10	20
FFT size (N_{FFT})	256	128	512	1024	2048
Number of used subcarriers (data and pilot)	200	84	420	840	1680
Number of guard band subcarriers and DC	56	44	92	184	368
Subcarrier frequency spacing, Δf (kHz)	15.625		10.94		
Useful symbol time, $Tb = 1/\Delta f$ (μs)	64		91.4		
Guard time assuming 12.5%, $Tg = Tb \times G$ (μs)	8		11.4		
OFDM symbol duration, $Ts = Tb + Tg$ (μs)	72		102.9		
Number of OFDM symbols in frame of 5 ms	69		48.0		
Ratio of cyclic prefix to useful symbol time (G)		1/32, 1/16, 1/8, 1/4			
Sampling factor (n)		Depends on used bandwidth: 7/6 for 256 OFDM, 8/7 for multiples of 1.75 MHz, and 28/25 for multiples of 1.25 MHz, 1.5, 2, or 2.75 MHz			

3.3 WiMAX Frame Structure

The frame in WiMAX can be represented as the frequency-division duplexing (FDD) mode or time-division duplexing (TDD) mode [12]. In the FDD mode, the transmission (DL) and reception (UL) of the data streams are simultaneously carried out in a given system bandwidth, but over separate frequencies [13]. On the other hand, in the TDD mode the transmission (DL) and reception (UL) of the data streams share the same bandwidth, but at a different time [14]. The WiMAX TDD frame is divided into two subframes [15]: the DL subframe and the UL subframe, as shown in Figure 3.2.

The transmission direction is switched in time by two time intervals, called transmit transition gap (TTG) and receive transition gap (RTG) [4]. The TTG is used to enable the base station to switch from the transmit to the receive mode, whereas the RTG is used to enable the base station to switch from the receive to the transmit mode [4]. These time intervals help to prevent the UL and DL transmission collisions [14].

The number of OFDM symbols in the TDD WiMAX frame depends on the used system bandwidth. In a 10 MHz system bandwidth, there are 48 OFDM symbols in a frame of 5 ms, and one of these symbols is used for the two gap types, TTG and RTG [16]. In order to divide the rest of the 47 OFDM symbols between the DL and UL subframes, the DL/UL ratio must be specified. To do so, the number of overhead symbols needs to be considered. The DL subframe begins with four control blocks (overheads) used by the base station to manage the communications in the network: preamble, frame control header (FCH), DL Map (DL-MAP), and UL Map (UL-MAP) [4]. These control blocks can be explained as follows.

First, the preamble is used for frequency and time synchronization and initial channel estimation, as well as noise and interference estimation [17].

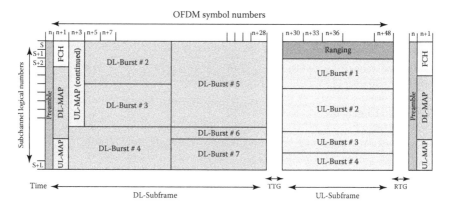

FIGURE 3.2
TDD WiMAX frame structure.

Second, the FCH holds frame control information, such as the DL-MAP message length (location) and used subchannels (in the case of segmentation); this information is declared by the DL_Frame_ Prefix message included in the FCH [17]. The third and fourth blocks are DL and UL MAPs. The DL and UL MAPs are MAC management messages that provide broadcast information about the burst locations in the DL and UL subframes [16]. For the DL-MAP, the burst start time is indicated by an information element (IE) message (DL-MAP_IE), which is declared in every DL-MAP [16]. In addition, the DL and UL MAPs refer to the burst profile type for both DL and UL users. The burst profile contains the modulation and forward error correction (FEC) [18]. However, the number of OFDM symbols reserved for DL and UL MAPS is variable, where both MAPs consist of fixed and variable parts. The variable part has the greatest impact in terms of increasing the MAP message length, where it depends on the number of users in the frame [19]. Continuously, the total number of overhead symbols depends on the number of OFDM symbol spans by the preamble, FCH, and DL and UL MAPs. The first OFDM symbol is used by the preamble, while the FCH and DL and UL MAPs should occupy an even number of OFDM symbols, as shown in Figure 3.2.

The WiMAX standard defines the slot as the smallest allocation data unit; the slot is defined by two dimensions: time and frequency [3]. Each slot comprises one subchannel over one, two, or more OFDM symbols. The number of OFDM symbols per slot depends on the link direction (DL or UL) and the type of subcarrier permutation. An adjacent set of slots assigned to a given user is called the user data region, and a group of data regions sharing the same channel condition is defined as a burst [8]. The burst contains data for one or more users, and each burst must be defined by one type of burst profile [16]. On the UL side, the subframe contains several bursts that are used to hold user data. The ranging part in the UL subframe is used for different purposes, such as network entry, power control, and bandwidth request. The DL and UL data bursts are illustrated in Figure 3.2.

3.4 Subcarrier Permutation Types

The WiMAX standard allows two types of subcarrier allocation into the subchannels in order to form logical resource allocation units (slots). These two types are used to satisfy user requirements under different channel conditions, which are discussed in the next sections.

3.4.1 Adjacent Subcarrier Permutation

Adjacent subcarrier means that the subcarriers are allocated into subchannels in sequence order [3]. Figure 3.3 shows an example of adjacent subcarrier permutation.

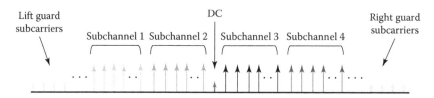

FIGURE 3.3
Adjacent subcarrier permutation.

In Figure 3.3, each set of arrows with the same color represents a group of adjacent subcarriers allocated to one subchannel. Consequently, the base station can select a suitable set of subcarriers according to the user channel condition, which enables the system to exploit multiuser diversity.

Band adaptive modulation and coding (band-AMC) is a type of adjacent subcarrier permutation mode [17]. In band-AMC, the data and pilot subcarriers are assigned to fixed locations that enable the base station to support the adaptive antenna system (AAS). The band-AMC uses bins to allocate subcarriers into slots: the bin contains eight data subcarriers and 1 pilot [3]. The slot is defined as $N_{bin}XM_{OFDM}=6$ [20] so that the slot can carry 24, 16, or 8 data subcarriers by 2, 3, or 6 OFDM symbols, respectively. The adjacent subcarrier mode is more suitable for use in fixed and low-mobility applications [21]. The number of subchannels in band-AMC mode is 48; therefore, the base station can allocate subchannels to users based on their frequency response.

3.4.2 Distributed Subcarrier Permutation

Distributed subcarrier means that the subcarriers are allocated into subchannels in a random way, thus enabling the system to exploit frequency diversity. In other words, different tones of frequency (subcarriers) are allocated in the subchannels; therefore, it is suitable for mobile applications [20]. Figure 3.4 shows an example of distributed subcarrier permutation.

The subchannels include a collection of subcarrier tones from high to low frequencies. One advantage of randomly allocated subcarriers is that they reduce the likelihood of subcarrier collisions with neighboring cells [22].

Partial Usage of Subchannels (PUSC) and Full Usage of Subchannels (FUSC) are two types of distributed subcarrier permutation modes. In PUSC, the slot

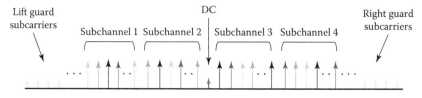

FIGURE 3.4
Distributed subcarrier permutation.

in the DL subframe spans two OFDM symbols, while in the UL subframe it spans three OFDM symbols [17]. Therefore, when PUSC is used, the ratio of DL/UL OFDM symbols must be a multiple of 2 in the DL direction and a multiple of 3 in the UL direction. The number of data subcarriers in the DL subframe under PUSC is 24 in one subchannel over one OFDM symbol and 48 in one subchannel over two OFDM symbols. In contrast, the number of subcarriers in the DL subframe under FUSC is 48 in one subchannel over one OFDM symbol. Thus, the number of subchannels in FUSC is reduced to half of that in PUSC.

The main design parameters of PUSC, FUSC, and band-AMC are listed in Table 3.2. These parameters correspond to a 10 MHz system bandwidth [20].

Analyzing Table 3.2, the number of subchannels (subcarriers) is different from one permutation to another. For instance, the band-AMC has the highest number of subchannels; hence, it offers more choices of different sets of subcarriers to be used by different user channel conditions (user diversity). On the other hand, FUSC has the lowest number of subchannels and holds all the slot subcarriers (48) in one OFDM symbol time, unlike PUSC, where the slot subcarriers are distributed through two OFDM symbol times for better channel tracking, where the subcarriers are randomly allocated per subchannel (frequency diversity). Therefore, the subcarrier ratio of the pilot to data in PUSC (0.167) is higher than that in FUSC (0.106) and band-AMC (0.125).

TABLE 3.2

Design Parameters of PUSC, FUSC, and Band-AMC

Subcarrier Permutation Type	Parameter	Value
PUSC	No. of subchannels	30
	No. of subchannels per major group	6
	No. of subchannels per minor group	4
	No. of PUSC groups	6
	No. of total data subcarriers	720
	No. of total pilot subcarriers	120
	No. of available slots in DL (for 30 OFDM symbols in DL)	450
FUSC	No. of subchannels	16
	No. of total data subcarriers	768
	No. of total pilot subcarriers	82
	No. of available slots in DL (for 30 OFDM symbols in DL)	480
Band-AMC (for $N_{bin}=2$, $M_{OFDM}=3$)	No. of subchannels	48
	No. of total data subcarriers	768
	No. of total pilot subcarriers	96
	No. of available slots in DL (for 30 OFDM symbols in DL)	480

The PUSC is distinct from the rest of the permutation types, due to the fact that it supports segmentation since it is able to form six groups of subchannels, defined as three major groups and three minor groups (Table 3.2), and these six groups can be divided into three segments [2]. This feature can be exploited in network deployment, specifically to control the interference effect (intercell interference [ICI]). For example, if the base station uses three sectors (three segments), then each of the two groups can be used by each sector. As a result, a frequency reuse factor of 1/3 is applied and the ICI is mitigated [22]. Segmentation is one of OFDMA's WiMAX features, and it is very useful when the available bandwidth is scarce or does not allow the use of a frequency reuse factor of more than 1— in other words, when the available bandwidth is not sufficient to be divided between cells.

3.5 Zone Concept in WiMAX Frame

The two types of subcarrier permutation represent two types of subcarrier allocation methods. The subcarriers are randomly allocated in PUSC, where they are sequentially allocated in band-AMC. Therefore, these two types of permutation cannot be used at the same time (same OFDM symbol time). A group of OFDM symbols sharing the same permutation type are gathered together to form a logical zone, or so-called permutation zone [23]. The DL and UL subframes may contain more than one zone, which enables the base station to serve different user types with different channel conditions. The starting and ending points of the zone boundaries are declared in the DL-MAP and UL-MAP for both DL and UL users, respectively [2]. On the DL side, a maximum of eight zones are allowed to be formed; the IE indicates the switching points between these zones through the STC_DL_Zone IE message, as shown in Figure 3.5 [4].

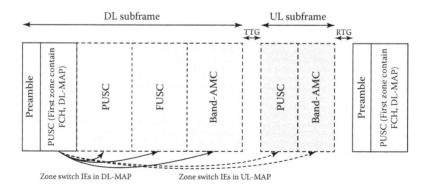

FIGURE 3.5
Zoning in TDD WiMAX frame.

The (mandatory) zones that must be used are indicated with a solid line, while the dashed line indicates the optional zones in the frame [2]. The optional zones may not be used in each frame; the use of these zones depends on the user application types. The PUSC should be used at the beginning of the frame to carry the control messages [23].

3.6 Modulation and Coding in Mobile WiMAX

One of the challenges in wireless communication is the variation in channel conditions during transmission time. An effective communications system must be able to respond to variations in channel condition in order to reduce the system error rate. WiMAX overcomes this challenge by supporting adaptive modulation. The adaptive modulation system uses less robust modulation types when the quality of the channel is high. In contrast, the system uses more robust modulation types when the quality of the channel is low [24].

In addition, WiMAX incorporates error correction techniques such as FEC to improve system throughput. The FEC helps to correct the errors at the receiver side by adding redundancy information to the transmitted signal. In spite of the fact that the redundancy reduces the amount of actual data in the transmitted signals, it is efficient in reducing the number of resend signals [24]. The FEC uses convolutional coding (CC), which is mandatory in the WiMAX system [3].

The WiMAX standard defines numerous burst profile types, where each burst profile contains information such as the modulation and coding scheme (MCS) type [3]. These burst profiles can be used by both base stations and users to adapt the transmission link to the channel condition. The recommended modulation and code types are listed in Table 3.3 [25]. Quadrature amplitude modulation (QAM) and quadrature phase-shift keying (QPSK) can be used in the UL and DL. However, CC is mandatory, while the convolutional turbo codes (CTCs) are optional, as well as the 64 QAM with a 5/6 code rate in the UL direction [9].

The base station specifies the suitable burst profile type for both connection directions (DL and UL), and informs users by two types of control messages: Downlink Interval Usage Code (DIUC) for DL users and Uplink Interval

TABLE 3.3

Recommended Modulation and Coding Types

Link Direction	Modulation Type	CC	CTC
DL	QPSK, 16 QAM, 64 QAM	1/2, 2/3, 3/4, 5/6	1/2, 2/3, 3/4, 5/6
UL	QPSK, 16 QAM, 64 QAM	1/2, 2/3, 5/6	1/2, 2/3, 5/6

Usage Code (UIUC) for UL users [4]. The DIUC and UIUC are included by IE in the DL and UL MAPs [16].

The base station alters the type of burst profile continuously according to the users' SINR to maintain connection quality. The base station requests the user to send the SINR by a report request (REP-REQ) MAC message, and the user responds by sending a report response (REP-RSP) MAC message. The SINR is reported in decibel scale [8]. There are two SINR threshold values defined for each burst profile: DIUC mandatory exit threshold and DIUC mandatory entry threshold [16]. These thresholds are used to select a suitable burst profile type. For DL, the procedures are as follows [16]:

If $SINR \leq DIUC$, mandatory exit threshold

$\rightarrow$ this DIUC can no longer be used

$\rightarrow$ change to a more robust DIUC

If $SINR \geq DIUC$, minimum entry threshold

$\rightarrow$ this DIUC can be used

$\rightarrow$ change to a less robust DIUC

If the received SINR value is above or below the predefined thresholds, the base station changes the burst profile type according to the new SINR level.

3.7 Scheduling and Types of Services Over WiMAX

The WiMAX MAC layer was constructed to support a variety of services to satisfy user demands. In general, the QoS is the process of delivering convincing services for different application requirements by ensuring the required QoS parameters for each application type (or service type). The WiMAX MAC layer can deal with different user applications, where five scheduling service (QoS classes) types are defined [19]. Each class of QoS has its own QoS parameters that satisfy the intended application requirements, such as [26] maximum and minimum throughput, traffic priority, maximum latency, and tolerated jitter. In order to enable the base station to schedule the requested service of a particular application, a service should be defined by maximum and minimum throughput that satisfy its QoS. The QoS parameters give the base station the flexibility to allocate resources (or bandwidth), and the priority which determines whether this service needs urgent delivery. For instance, video as a real-time application has a higher priority of bandwidth allocation than File Transfer Protocol (FTP) and e-mail. Besides, the allowable delay in data transmission (latency) from the sender to the receiver should be defined, and the delay variation (jitter) during transmission time should be considered as well. The five QoS classes that are defined by the WiMAX standard are as follows [19]:

1. *Unsolicited Grant Service (UGS)*: The UGS is created to support real-time data applications that require fixed-size data packets issued at constant (periodic) intervals, such as the Voice over Internet Protocol (VOIP) without silence suppression. VOIP often has a talk-spurt period (sometimes called an active period) followed by a silent period. The base station provides fixed-size data grants at periodic intervals, which override the overhead of user bandwidth requests. Once the connection is established, there is no need to send any other requests.

2. *Real-Time Polling Service (rtPS)*: rtPS is designed to support real-time data applications comprised of variable-size data packets that are issued at constant (periodic) intervals, such as Moving Pictures Experts Group (MPEG) video transmission. In rtPS, the base station provides periodic UL request (unicast) opportunities for the user, which allows the user to specify the need of the desired grant size (or bandwidth). This service needs more request overhead than UGS, since it supports variable grant sizes based on user demand.

3. *Extended Real-Time Polling Service (ertPS)*: ertPS is appropriate for variable-rate real-time applications that require a specific data rate and delay, such as VOIP with silence suppression; no traffic is sent during silent periods. This service is built on the efficiency of both UGS and rtPS. The ertPS provides unicast grants like UGS, so it reduces the overhead of the bandwidth request from the user and supports variable packet sizes, like rtPS.

4. *Non-Real-Time Polling Service (nrtPS)*: nrtPS is suitable for delay-tolerant data streams such as FTP transmission, which consist of variable-size data packets that require a minimum data rate transmitted at regular time intervals (not necessarily periodic time intervals). For bandwidth request, the user may use contention request opportunities, as well as unicast request opportunities.

5. *Best-effort (BE) service*: BE is designed to support data streams such as web browsing using the Hypertext Transfer Protocol (HTTP), which does not require a minimum service level guarantee, and therefore it can be handled on a best-available basis. If there is bandwidth left from other service classes, then the BE service grants bandwidth to the user.

UGS, rtPS, nrtPS, and BE were originally available in 802.16–2004 WiMAX, whereas the ertPS service is joined by 802.16e–2005 WiMAX.

In order to schedule these five QoS classes, the base station schedule in the MAC layer should be involved. In PMP, the base stations use a centralized MAC architecture to carry out the required tasks, such as determining UL and DL access [16]. The base station scheduler performs complex procedures to determine the required bandwidth and burst profile type for each connection, which depends on the link quality, network load, and service type. Resource allocation in OFDMA can be partially implemented, since the resources are defined by two dimensions: time and frequency, which increase the complexity of the base station scheduler [19]. However, this results in an increase in bandwidth (or resource) utilization [19]. In order to make the appropriate decision, the base station scheduler needs to analyze many entry parameters, such as the received SINR value of the involved link, the amount of requested bandwidth, and the required QoS class with its parameters to ensure the quality of the connection. This can be seen in Figure 3.6, where the base station scheduler operation is illustrated [16].

The base station scheduler must permanently monitor the received SINR value of the involved links. Based on the SINR value, the modulation and coding type are determined (burst profile type), which in turn helps to specify the amount of resources (slots in time and subcarriers in frequency) that should take place in the intended transmission opportunity. This should be in line with the service class type and its associated parameters. Hence, the parameters of each service class type show the required throughput and latency. The base station can then explicitly determine the suitable modulation type and the period in which these resources should be scheduled during the transmission opportunities.

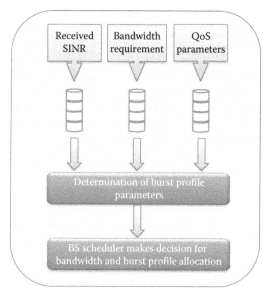

FIGURE 3.6
WiMAX base station scheduler operation.

The base station scheduler handles the UL request by decision-making for UL resource allocations, and these allocations are sent to the beneficiaries through the UL-MAP MAC management message. Similarly, the base station scheduler takes care of the DL resource request and informs the beneficiaries about the locations of these sources in the DL-MAP MAC management message. When the TDD frame mode is used, the transmission time is divided between the UL and DL directions; thus, all the base station schedulers in the network can operate simultaneously to allocate resources [16]. However, modifying the operation of the base station scheduler can be beneficial in several ways, such as eliminating ICI by controlling the resource allocation in time and frequency or by inserting certain rules to allocate resources based of the SINR values, such as is the case in the FFR technique. Importantly, the WiMAX standard does not specify any specific algorithm for the base station scheduler; any popular algorithms can be used, such as [16] round-robin, weighted round-robin, and weighted fair queuing. Vendors may choose any of them based on their requirements, such as high efficiency and easy-to-manage resources.

3.8 WiMAX Base Station Deployment Using the FFR Technique

The many advantages of WiMAX technology make the WiMAX base station flexible enough to be suitable for several types of network deployment environments. Wireless networks can be deployed in rural, subrural, and urban environments, where each has specific requirements to work properly [22]. WiMAX as a technology supports several features, of which the most important are different subcarrier allocation modes, zones forming, segmentation capabilities, and adaptive modulation. These features can be considered great tools to deploy WiMAX base stations in different environments.

WiMAX base stations are deployed in cellular networks where different frequency reuse factors can be used. If the reuse factor equals 1, then all the bandwidth is used by each cell. In contrast, if the reuse factor equals 3, then one-third of the bandwidth is used in each cell. In the former scheme, high spectral efficiency can be achieved, but with degradation in the quality of service for cell edge users as a result of the effect of interference. On the other hand, spectral efficiency is less in the latter scheme because the bandwidth is divided by 3, which results in low interference (ICI). FFR can be considered as a compromise to simultaneously improve the spectral efficiency and reduce the ICI effect. Figure 3.7 shows FFR WiMAX base station deployment.

The FFR technique plays an important role in cellular network deployment, as it supports universal frequency reuse with a slight reduction in spectral efficiency [9]. FFR takes advantage of zoning and segmentation, as well as having the ability to be implemented by different subcarrier permutation types. The FFR technique is verified by the WiMAX Forum [9,27] to satisfy the connection requirements of cell border users in cellular network deployment without invoking the frequency planning technique. FFR aims to provide full subchannel usage for users near the base station, and with the help of the PUSC mode, it provides a fraction of subchannels for users far away from the base station [9].

The DL subframes of three adjacent cells are depicted in Figure 3.7a, and the frequency mapping of these DL subframes is illustrated in Figure 3.7b [9]. Each DL subframe in Figure 3.7 has two zones to serve inner and outer users, namely, R1 and R3, respectively. In the R3 zone, the available bandwidth is divided into three equal segments, A, B, and C, where each segment uses part of the bandwidth F1, F2, and F3, respectively, to form three parts of orthogonal subcarriers, as in Equation 3.1. This means that these subcarriers in the three segments are orthogonal to each other and can be used in each cell border to reduce the interference between adjacent cells.

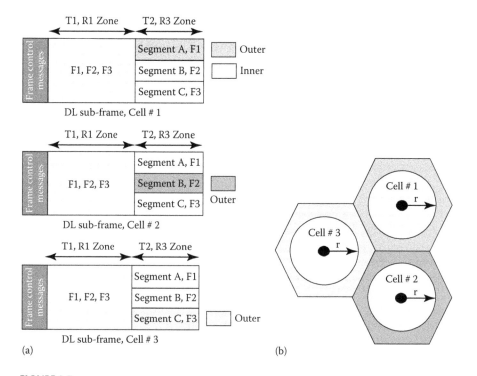

FIGURE 3.7

WiMAX base station deployment using the FFR technique. (a) DL subframe structures. (b) Frequency mapping per cell.

$$BW = F1 \cup F2 \cup F3$$

$$\text{Where } |Fi| = \frac{1}{3}BW$$

$$\text{Then } Fi \cap Fj \cap Fk = \varnothing$$

$$\text{Such that } i \neq j \neq k \text{ for } i, j, k \in \{1, 2, 3\}$$

$$(3.1)$$

The concept behind using FFR is to enhance the signal power of cell border users. Therefore, there is a need to use a criterion such as SINR threshold to keep the user signal strength at an acceptable level. The operation time of the DL subframe is divided into two parts corresponding to the available number of OFDM symbols in the R1 and R3 zones, as shown in Figure 3.7a. In the R1 zone, all the available subchannels (subcarriers) are used; hence, an increase in spectral efficiency can be considered a compensation for using part of the subchannels in the R3 zone. In the R3 zone, only part of the subchannels (one-third) are used to serve cell border users. The inefficient utilization of bandwidth in the R3 zone is considered a weak point in the FFR technique. However, this is necessary to mitigate the interference between adjacent cells. On the other hand, if all the available subchannels in the R3 zone can be utilized, then the FFR technique can achieve a frequency reuse of 1, which significantly enhances FFR performance in terms of spectral efficiency and data rate.

In R1 operation time (T1), users in the cell center area receive useful power from their base station and interference power from other base stations that are using the same frequency band at the same time. The received power decays as a function of increasing the distance between the sender and receiver. Therefore, the freedom of movement for the cell center users is limited to a certain distance (or radius r) from the base station and is offset by a suitable SINR threshold value, as shown in Figure 3.7b.

In order to choose the SINR threshold, it is better to choose a value in which the signal becomes weak. Most of the works done in this area consider a low SINR value [28,29]. Therefore, to save the user signal strength, serve them in the R3 zone, since R3 zone users suffer less interference. In R3 zone operation time (T2), the neighboring base stations use orthogonal subchannels in their cell borders. Thus, these subchannels will not interfere with each other, and that will increase the SINR value of the cell border users.

In terms of design, two purposes can be obtained from the SINR threshold. First, it specifies the cell center radius, and second, it separates the inner and outer users. The standards do not provide a lot of technical details about the FFR technique. The design stages of the FFR technique, depicted in Figure 3.7, can be summarized as follows:

1. Specify a reasonable number of OFDM symbols for the R1 and R3 zones. These numbers should match the slot definition for a particular subcarrier permutation type.

2. The number of OFDM symbols in the R1 or R3 zone represents the amount of resources (slots) that will be reserved for user data. Since the FFR does not use all subchannels in the R3 zone, it is advised to use it in low-population environments, such as a rural area [30]. Therefore, it is better to choose a higher number of OFDM symbols in the R1 zone than that in R3 zone, where the high population density is concentrated near the base station.

3. Specify the SINR threshold ($SINR_{TH}$) to determine the radiuses of the cell center area and cell border area, as well as to distinguish between the inner and outer users or mobile stations (MSs). The resource assignment in the FFR technique based on SINR can be presented as follows:

FFR RESOURCE ASSIGNMENT ALGORITHM

BEGIN

1-   **for** i=1 **to** N_{MS} **do**   /* N_{MS} *is number of users need service* */

2-   **Compute** *SINR(MSi)*

3-   **Zone = 0**

4-   **if** *SINR(MSi)* ≥ *SINR$_{TH}$*   **then**

5-   **Zone** = R1   /*define R1 Zone as an operational Zone */

6-   R1 ⟵ MSi   /*Assign user to R1 Zone*/

7-   MSi ⟵ BPj, : BPj ≡ SINR(MSi) /* Specify suitable burst profile*/

8-   **else**

9-   **if** *SINR(MSi)* < *SINR$_{TH}$*   **then**

10-   **Zone** = R3   /*define R3 Zone as an operational Zone */

11-   R3 ⟵ MSi   /* Assign user to R3 Zone*/

12-   **MSi** ⟵ **BPj,** : BPj ≡ SINR(MSi) /* Specify suitable burst profile*/

13-   **endif**

14-   **endif**

15- **endfor**

END

The resources (slots) are assigned to either R1 or R3 zones according to the SINR threshold. However, when the zone is determined, the base station will then choose a suitable burst profile (BP) based on the reported SINR (SINR(MS)), which depends on the channel condition of the intended user.

3.9 Summary

In this chapter, the most important materials and features of the WiMAX base station related to the contribution of this book have been presented and discussed. The WiMAX base station supports different user types, such as fixed, nomadic, and mobile. Therefore, strong design tools are required to make this technology compatible with the advances in the world of communications.

The WiMAX frame can be divided into DL and UL subframes. The DL subframe is used to deliver data to users, while the UL subframe is used to transfer user data to another entity. The DL and UL subframes can be divided into several zones, and in each zone one type of subcarrier permutation can be used.

Subcarrier permutation is a method that can be manipulated to serve different user types. For example, adjacent subcarrier permutation enables the base station to achieve user diversity, as is the case in band-AMC mode. On the other hand, distributed subcarrier permutation enables the base station to achieve frequency diversity, as is the case in PUSC and FUSC. These two types of subcarrier permutation permit the base station to adapt to different environments and different user types.

In addition, the WiMAX frame supports segmentation, where the zone itself can be further subdivided into several segments. Each segment may use one or more subchannel groups. The standard allows for six groups when PUSC is involved. One or more of these groups can be used to serve part of the cell coverage area, which is the case of the FFR technique.

WiMAX base stations can be deployed in cellular networks with universal frequency, but at the cost of losing the connection quality of cell border users, which is not the case when FFR is considered. When FFR is applied, universal frequency can be implemented in the cell center area while ensuring the connection quality of cell border users. FFR can be considered a trade-off between spectral efficiency and ICI. FFR uses the full bandwidth to serve users near the base station and a fraction of the bandwidth to serve users far away from the base station.

One of the advantages of using FFR is the signal recovery of cell border users, since interference is reduced. However, one of the disadvantages is using a portion of the bandwidth in the R3 zone, which results in inefficient bandwidth and resource utilization. In Chapter 4, the methodology of modeling a new FFR technique based on the WiMAX cellular network will be presented.

References

1. Y. Zhang, *WiMAX Network Planning and Optimization*, Auerbach Publications, Boca Raton, FL, 2009.
2. Forum, WiMAX™ system evaluation methodology, version 2.1, 2008, p. 209.
3. J. G. Andrews, A. Ghosh, and R. Muhamed, *Fundamentals of WiMAX: Understanding Broadband Wireless Networking*, Pearson Education, London, 2007.
4. IEEE-Std, IEEE 802.16–2009: IEEE standard for local and metropolitan area networks part 16: Air interface for broadband wireless access systems, in IEEE Standards (Revision of IEEE Standards 802.16–2004), 2009, pp. 1–2080.
5. M. Ergen, *Mobile Broadband: Including WiMAX and LTE*, Springer, Berlin, 2009.
6. H. Rohling, *OFDM: Concepts for Future Communication Systems*, Springer, Berlin, 2011.
7. D. Sweeney, *WiMax Operator's Manual: Building 802.16 Wireless Networks*, Dreamtech Press, India, 2007.
8. IEEE-Std, IEEE 802.16e–2005: IEEE standard for local and metropolitan area networks part 16: Air interface for fixed and mobile broadband wireless access systems amendment for physical and medium access control layers for combined fixed and mobile operation in licensed bands, in IEEE Standards, 2005, ieeexplore.ieee.org/document/1603394/, 2005.
9. Forum, Mobile WiMAX—Part I: A technical overview and performance evaluation, 2006, p. 53.
10. I. Akyildiz and X. Wang, *Wireless Mesh Networks*, vol. 3, Wiley, Hoboken, NJ, 2009.
11. F. Ohrtman, *WiMAX Handbook*, McGraw-Hill, New York, 2005.
12. W. Roh and V. Yanover, Introduction to WiMAX technology, in *WiMAX Evolution: Emerging Technologies and Applications*, ed. M. D. Katz and F. H. Fitzek, Wiley, Hoboken, NJ, 2009, pp. 1–13.
13. M. D. Katz and F. H. Fitzek, *WiMAX Evolution*, Wiley, Hoboken, NJ, 2009.
14. G. Kar, Performance of static and adaptive subchannel allocation schemes for fractional frequency reuse in Wimax networks, Master dissertation, Bilkent Universty, 2011.
15. G. Radha, S. Krishna, V. Rao, and G. Radhamani, WiMAX: A wireless technology revolution, Auerbach Publications, Taylor & Francis Group, Boca Raton, FL, 2008.
16. L. Nuaymi, *WiMAX: Technology for Broadband Wireless Access*, Wiley, Hoboken, NJ, 2007.
17. K. Balachandran, D. Calin, F. C. Cheng, N. Joshi, J. H. Kang, A. Kogiantis, et al., Design and analysis of an IEEE 802.16 e-based OFDMA communication system, *Bell Labs Technical Journal*, vol. 11, pp. 53–73, 2007.
18. M. C. Necker, A novel algorithm for distributed dynamic interference coordination in cellular OFDMA networks, Doctor-engineer, Faculty of Computer Science, Electrical Engineering and Information Technology, Stuttgart, 2009, vesta.informatik.rwth-aachen.de/opus/volltexte/2011/2980/pdf/27.pdf.
19. C. So-In, R. Jain, and A.-K. Tamimi, Capacity evaluation for IEEE 802.16 e mobile WiMAX, *Journal of Computer Systems, Networks, and Communications*, vol. 2010, p. 1, 2010.

20. U. D. Dalal and Y. P. Kosta, *WIMAX, New Developments*, In-Tech, Rijeka, Croatia, 2009.
21. M. Salman, R. Ahmad, M. K. Sharief, and M. S. Al-Janabi, Simulation study of WiMAX base station deployment using AMC under different frequency planning techniques, in *2014 2nd International Conference on Electronic Design (ICED)*, 2014, pp. 537–542.
22. L. Korowajczuk, *LTE, WIMAX and WLAN Network Design, Optimization and Performance Analysis*, Wiley, Hoboken, NJ, 2011.
23. M. Riegel, A. Chindapol, and D. Kroeselberg, *Deploying Mobile WiMAX*, Wiley, Hoboken, NJ, 2010.
24. R. Prasad and F. J. Velez, *WiMAX Networks: Techno-Economic Vision and Challenges*, Springer, Berlin, 2010.
25. A. Kumar, *Mobile Broadcasting with WiMAX: Principles, Technology, and Applications*, Taylor & Francis US, Boca Raton, FL, 2008.
26. Y. Zhang and H.-H. Chen, *Mobile WiMAX: Toward Broadband Wireless Metropolitan Area Networks*, CRC Press, Boca Raton, FL, 2007.
27. Forum, Mobile WiMAX—Part II: A comparative analysis, 2006, p. 47.
28. S. R. Boddu, A. Mukhopadhyay, B. V. Philip, and S. S. Das, Bandwidth partitioning and SINR threshold design analysis of fractional frequency reuse, in *National Conference on Communications (NCC)*, New Delhi, India, 2013, pp. 1–5.
29. Y. Zhou and N. Zein, Simulation study of fractional frequency reuse for mobile WiMAX, in *IEEE Vehicular Technology Conference, VTC Spring 2008*, Canada, 2008, pp. 2592–2595.
30. R. Giuliano, C. Monti, and P. Loreti, WiMAX fractional frequency reuse for rural environments, *IEEE Wireless Communications*, vol. 15, pp. 60–65, 2008.

4

Design and Modeling of Resource Utilization in Fractional Frequency Reuse

4.1 Introduction

The demand for mobile wireless communication has dramatically increased with the emergence of modern applications that support daily needs, such as e-mail and multimedia. Cellular networks are used to increase the coverage area of wireless communications, where base stations are deployed to adjust each other's connection to increase the service coverage area. One of the drawbacks in cellular networks is intercell interference (ICI), which is caused by using similar frequency bands in neighboring cells.

Many solutions have been proposed to overcome the ICI problem, like frequency planning, smart antenna [1], and traditional fractional frequency reuse (FFR) [2]. Traditional FFR solves the problem of ICI efficiently and enhances signal quality, especially for cell edge users. However, the traditional FFR approach is the inefficient utilization of resources and bandwidth, which cause a reduction in spectral efficiency, data rate, and number of served users and a shortage in resources.

In this book, two new algorithms are proposed, designed, and modeled, called static resource assignment (SRA) and dynamic resource assignment (DRA), to enhance traditional FFR performance. A summary of the work done in this chapter is presented in Figure 4.1 as a block diagram.

The block diagram starts by analyzing the FFR downlink (DL) subframe parts in an interference environment in order to study the possibility of using these parts more effectively. In addition, it is used to specify the service area of each part in the DL subframe. The SRA algorithm is utilized to tackle the deficit in radio resources of the traditional FFR technique, where four situations or cases are considered as a trade-off study to find the best way to enhance the performance of traditional FFR. Although SRA enhances the performance of traditional FFR in terms of number of served users, resource utilization, data rate, and spectral efficiency, it is not able to address the problem of variation in population density. This is because the density of mobile users is subject to continuous change. Therefore, the DRA algorithm is proposed to address this problem, in addition to enhancing the performance of traditional FFR in terms of the same metrics that are used in SRA FFR. In the DRA algorithm, two types of user distribution are proposed to explore the performance of this algorithm in different types of mobility patterns. It is worth mentioning that the SRA mode is closer to the static configuration mode, where the system parameters are set earlier and remain unchanged for a period of time. In contrast, the DRA algorithm is closer to the dynamic configuration mode, where the system parameters are set dynamically frame by frame according to the cell load.

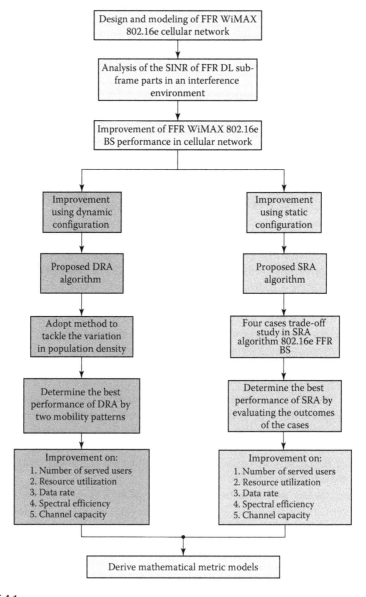

FIGURE 4.1
Flow of research works to improve performance of FFR WiMAX 802.16e base station in cellular network.

The rest of this chapter is organized as follows: In Section 4.2, the network environment and system parameters are depicted, along with the new DL subframe structure of the proposed models (SRA and DRA). Section 4.2 also provides extensive analysis of the DL subframe parts in an interference environment. The resource assignment scenario and the algorithm steps of the proposed models are discussed in Sections 4.3 and 4.4, respectively. The system modeling is presented in Section 4.5, followed by performance evaluation criteria in Section 4.6. The work done in this chapter is summarized in Section 4.7.

4.2 System Model Design

This section is divided into two parts. First, the network design and the required system parameters are introduced, which represents the environment of the new design. In addition, an interference model is presented by considering the signal-to-interference-plus-noise ratio (SINR) levels of all base stations in the grid. The objective of considering the interference between cells is to analyze the performance of traditional FFR DL subframe parts in an interference environment. This analysis helps us to use these parts more effectively. Second, details of the new DL subframe design are presented, along with the resource assignment scenario (slot) per user load. The MATLAB™ software used to simulate the work done in this book follows the 802.16e standard document described in the evaluation methodology of the WiMAX Forum [3].

4.2.1 Network Design and System Parameters

The considered network based on the proposed algorithms is a collection of base stations deployed adjacent to each other to form a cellular system of two tiers. Network size is 19 hexagonal cells, where each cell contains one base station and all these base stations transmit a fixed and equal power. The first tier consists of 6 cells surrounding the center cell, and the second tier consists of 12 cells surrounding the center cell. In a cellular system, the distribution of cells may take the shape of a hexagon, square, circle, or other regular shape [4]. The hexagon has been widely adopted in network simulations, since it allows easy and manageable analysis in a cellular system [5]. In the real world of network deployment, the cell shapes are irregular due to the differences in buildings, terrain, and other factors [6]. It is assumed that all the base stations in the network use the FFR technique to mitigate the ICI effect. The proposed network layout is illustrated in Figure 4.2.

FFR uses time-division duplexing (TDD) to manage the available resources (slots) between near and far users. The resulting DL subframes of three base

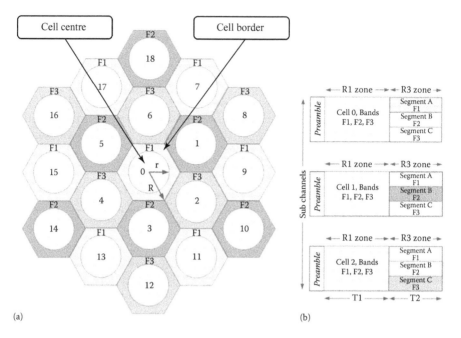

FIGURE 4.2
Proposed network layout. (a) Cell frequency distribution. (b) DL subframe structure.

stations are illustrated in Figure 4.2b. The frequency distributions for the 19 cells are depicted in Figure 4.2a, where every three adjacent cells that form a triangle use different frequency bands in their boundary (F1, F2, and F3). Each frequency band represents a group of subchannels, and these subchannels are orthogonal to each other. This frequency distribution is mandatory to ensure different types of frequency bands in each cell border, in order to avoid frequency interference between adjacent cells.

In the FFR technique, the DL subframe is divided into two periods of time slots: T1 to serve users in the center area and T2 to serve users in the border area. In T1, all the base stations use the available bandwidth (all subchannels) in the system, which leads to an increase in the ICI, especially at the cell border [2]. Therefore, there is a limitation to serve users in the cell center area, which is defined by the cell center radius (r). In contrast, users with low signal strength are served in the cell border area, where the interference is mitigated, and the limit to serve these users is defined by the cell radius (R) as depicted in cell number zero in Figure 4.2a. In network deployment, it is useful to use the full bandwidth (F1, F2, and F3) to increase spectral efficiency, but this is unapplicable because of the ICI effect. However, all the cells in the grid use universal frequency (FRF = 1) to serve users at the cell center area, whereas they use part of the bandwidth (FRF = 3) to serve users at the cell border. Note that it should not be confused with FFR and FRF; the former refers to a technique, whereas the latter refers to the frequency reuse factor. In FFR, each base station can use the available bandwidth in the system (universal frequency) except in the R3 zone, where one-third (either F1, F2, or F3) of the bandwidth is used to control the ICI [7]. That is why FFR is used in cellular systems despite the loss of two-thirds of the bandwidth in the R3 zone. According to the losses of two-thirds of bandwidth (or resources) in the R3 zone, FFR works well in areas with a low population density, such as a rural environment [8].

The work done in this book is intended to enhance the performance of the IEEE 802.16e–2005 WiMAX base station using the FFR technique by taking advantage of unused parts in the DL subframe—specifically by utilizing segments B and C (segment BC) in the R3 zone (see the DL subframe of cell zero in Figure 4.1b). In other words, all the available frequency bands (F1, F2, and F3, or equivalent segments A, B, and C) are utilized in the R3 zone. However, at T2 time slots of R3 zone, if all the subchannels in segments A, B, and C are used to serve cell border users at the same time, then the ICI is increased, which leads to dispelling the interest of using the FFR technique. The ICI is increased since the subchannels of segments A, B, and C are not orthogonal to the other subchannels of the surrounding cells in the grid, which prevents the use of segment BC. In order to avoid this, two conditions must be taken into consideration. First, segment BC serves users located at the cell center area, where the co-channel effect is far enough to avoid the effect of ICI. Second, determination of the service area of segment BC by defining the farthest point from the base station allows segment BC to serve users in the cell center area by keeping the signal strength of these users at an acceptable level. The second condition can be achieved by defining a suitable SINR threshold when making decisions to serve users in segment BC. One of the benefits of using SINR is that it takes into account the interference effect of neighboring cells when assigning resources. Equation 4.1 shows the basic formula to calculate SINR.

$$\text{SINR} = \frac{\text{Useful power}(i)}{\sum_{j=1}^{l} \text{Interference power}(j) + No} \tag{4.1}$$

The numerator of Equation 4.1 represents the power received by a user from his own serving base station i, and the denominator represents the interference power received by the same user from other base stations (j to l) that use the same frequency bands of the serving base station plus thermal noise power (No), where $j \neq i$. However, l represents the number of base stations that cause interference in the grid, which are denoted as co-channel cells (Figure 2.1). The value of l depends on the run time of orthogonal frequency-division multiplexing (OFDM) symbols in the DL subframe, which in turn depends on the operational time of the intended part (R1 and R3 zones, or segment BC) in the DL subframe. Each zone or segment has start and end points of OFDM symbols. If M denotes the target zone or segment and τ denotes the operation time of a given OFDM symbol in the DL subframe, then the value of l can be determined as follows:

$$l = \begin{cases} 18 := \tau \forall T1, & \text{when } M\{R1 \text{ zone}\} \\ 6 := \tau \forall T2, & \text{when } M\{R3 \text{ zone (segment A)}\} \\ 12 := \tau \forall T2, & \text{when } M\{\text{Segments BC}\} \end{cases} \quad (4.2)$$

Referring to Figure 4.2, if the R1 zone is under consideration, then l equals 18, where $l \in \{1, 2, ..., 18\}$, since all the base stations use all the available subchannels (F1, F2, and F3) in T1 time slots. In contrast, if segment A in the R3 zone is under consideration, then l equals 6, where $l \in \{7, 9, 11, 13, 15, 17\}$, since only six base stations use the same subchannels (F1) at T2 time slots. When considering segment BC, l equals 12, where $l \in \{1, 2, 3, 4, 5, 6, 8, 10, 12, 14, 16, 18\}$, since only 12 base stations use the same subchannels (F2 and F3) at T2 time slots. The subchannels of segments A (F1) are omitted since they are orthogonal to the subchannels of segment BC (F2 and F3) at T2 time slots. Using Equation 4.1, and adopting the same notation presented in [9,10], the average SINR versus distance for different values of l is plotted in Figure 4.3, when a user in cell zero moves on the X-axis from the base station toward the cell border. Responses of four SINR lines are plotted in Figure 4.3, which are traditional FFR (Trd. FFR), proposed FFR (Pro. FFR), FRF of 1, and segment BC (Seg. BC). The results obtained in Figure 4.3 allow

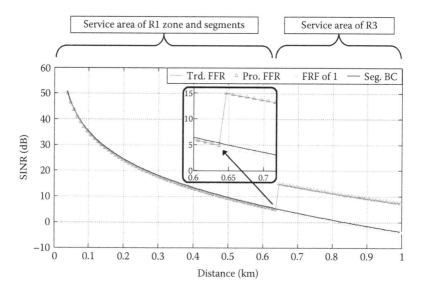

FIGURE 4.3
SINR versus distance from BS (when FFR 5 dB threshold is used).

the determination of the service area for each R1 zone (when $l=18$), R3 zone (when $l=6$), and segment BC (when $l=12$). The average SINR line of FRF of 1 (universal frequency) is added to show the worst interference case and to illustrate the benefit of using the FFR technique for cell edge users. Using universal frequency in network deployment affects the signal quality of cell border users, as revealed in Figure 4.3, where the SINR is decreased below 0 dB beyond 800 m.

Referring to Figure 4.3, the maximum allowable distance to serve users in the R1 zone is equal to 635 m, when the SINR threshold value of 5 dB is selected [11]. The selection of 5 dB is to maintain users' signal quality, since after 635 m (less than 5 dB) the options to use different modulation types will be less. The SINR value in the R3 zone is increased by about 10 dB as result of using the FFR technique, where the interference at the cell border is reduced ($l=6$). Importantly, the SINR line of segment BC is slightly increased compared with the SINR line of the R1 zone during T2, due to the exclusion of the six cells of segment (A). The SINR of segment BC is high enough to serve users in the cell center area, since the co-channel effect is mitigated as a result of large path loss. In return, segment BC can be used to serve users in the cell center area. Therefore, the base station at the time slot of T2 is able to serve users in the cell border (segment A) and cell center (segment BC). However, the SINR of the proposed FFR is identical to that of the traditional FFR, since using segment BC will not increase the interference level. Segment BC serves users in the cell center area, but with different time slots than that of R1 zone (see Equation 4.2). In other words, the utilization of segment BC is based on controlling the locations of users and the operational time of segment BC.

In reference to Figure 4.3, the SINR value of segment BC is equal to 5.38 dB at a distance of 635 m. In order to ensure best performance for users who use segment BC, the threshold of 5 dB is used to activate these segments. As a result, the service area of segment BC is specified to be between 36 and 635 m, the same as the boundaries of R1 zone. Thirty-six meters is the minimum distance allowed from the base station to serve users [3]. In conclusion, the service areas of the R1 and R3 zones are 36–635 m and 635–1000 m, respectively, whereas the service area of segment BC is from 36 to 635 m.

The proposed design parameters of the base station are listed in Table 4.1; most of these parameters are defined in the WiMAX Forum [3]. The specifications in Table 4.1 enable the base station to work properly in line-of-sight (LOS) and non-line-of-sight (NLOS) conditions to support mobile applications. Moreover, all these parameters are built on the basis of using an orthogonal frequency-division multiple access (OFDMA) IEEE 802.16e WiMAX frame duration of 5 ms with a 10 MHz bandwidth.

In order to enable the TDD mode to work properly, it is assumed that all the base stations are perfectly synchronized to keep unified frame start times and frame durations, as well as that all the base stations use unified

TABLE 4.1

Proposed Design Parameters

Item Description	Value
Number of cells per cluster	19
Operating frequency	2500 MHz
Bandwidth (BW)	10 MHz
Permutation mode	PUSC
FFT size (N_{FFT})	1024 subcarriers
Subcarrier frequency spacing Δf	10.94 kHz
Useful symbol time ($Tb = 1/\Delta f$)	91.4 µs
Guard time ($Tg = Tb \times G$)	11.43 µs
G, Ratio of cyclic prefix ($G = Tg/Tb$)	1/8, for 10 MHz BW
OFDMA symbol duration ($Ts = Tb + Tg$)	102.9 µs
Number of OFDM symbols (S_{frame})	48 (in frame of 5 ms)
Sampling factor (n)	28/25, for 10 MHz BW
Frame duration ($Tf = Ts \times S_{frame}$)	5 ms
Number of slots per two successive OFDM symbols (γ)	30
Number of subcarriers per slot (Kr_{slot})	48
Cell center area radius (r)	635 m
Cell border area radius (R)	1000 m
Traffic ratio (DL/UL)	29/18
User distribution	Random per drop
Duplexing mode	TDD
Adaptive modulation and coding	Enable
DL power control	Switched off

boundaries for the R1 and R3 zones to avoid interzone interference [8]. In addition, the connection topology between users and the base station is assumed to be point-to-multipoint (PMP), and all the base stations use the Partial Usage of Subchannels (PUSC) mode, where the mapping of the 1024 subcarriers in the DL subframe is configured as follows: 120 for pilots, 184 for guards, and 720 for data.

It is assumed that a simple scheduler is be used (first in, first out) [12]. In the scheduler, the allocation of user resources to a particular zone or segment is based on a predefined SINR threshold. The scheduler also considers user load type and the available free slots in the DL subframe. The links between the base station and users are assumed to be error-free and are adaptively modulated, where adaptive modulation and coding is considered in this design. The cell radius (r) is obtained based on analyzing the SINR level for the R1 zone and segment BC. The test period is set to 100 trials (or attempts) [13], where users are randomly distributed in the Cartesian coordinate system (X-axis and Y-axis). The locations of the users are random variables that follow the uniform distribution, and the random distribution of users gives the sense of mobility.

4.2.2 Proposed DL Subframe Structure

Subchannelization and zoning are two features that help to partition the WiMAX frame into different shapes, as mentioned in Sections 3.2 and 3.5, respectively. FFR implementation utilizes these features in WiMAX to build the DL subframe. The proposed DL subframe is shown in Figure 4.4.

The WiMAX Forum defines default values for the ratio of DL/UL OFDM symbols equal to $(47\text{-}N_{OFDM}^{UL}/N_{OFDM}^{UL})$, where the number of OFDM symbols in the UL is equal to $12 < N_{OFDM}^{UL} < 21$ for a WiMAX frame duration of 5 ms [3]. The number of OFDM symbols in the WiMAX frame is 48 [2], where one of these is used by transmit transition gap (TTG) and receive transition gap (RTG) (see Section 3.3), and the remaining 47 symbols are used as follows: if the overhead in the DL subframe is assumed to occupy the first 3 symbols, since assuming fixed overhead is acceptable, such as in [7,14], then 44 OFDM symbols are left in the WiMAX frame. If 18 OFDM symbols are used for the UL subframe, then (44 − 18) are the remaining OFDM symbols for the DL subframe; thus, the ratio of the DL/UL is assumed to be 29/18. As a result, if the R3 zone spans 10 symbols, then the R1 zone will span 16 symbols, as shown in Figure 4.4.

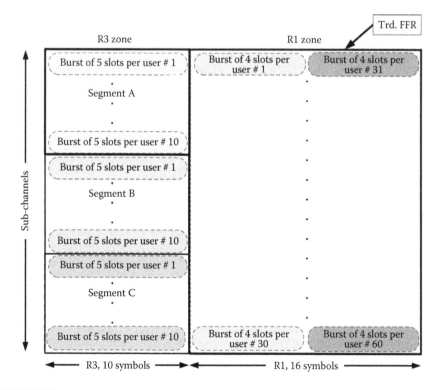

FIGURE 4.4
Proposed DL subframe without overhead.

The WiMAX Forum defines the slot, which is the smallest time-frequency data allocation unit, or so-called radio resource, that can accommodate user data, where the slot is defined by two dimensions: time (OFDM symbols) and frequency (subchannels) [3]. However, if the PUSC mode is used in the DL subframe, then the slot occupies two OFDM symbols in the time domain and one subchannel in the frequency domain [15]. In addition, a set of contiguous slots share the same channel condition, called data region or burst [15]. Consequently, the total number of slots in the DL subframe can be specified. However, knowing the total number of slots in the DL subframe and the number of slots required per user (user load) means that the total number of users that can be served in the DL subframe can be determined [16].

In reference to the proposed DL subframe in Figure 4.4, each row represents eight slots in the R1 zone, and each row of each segment in the R3 zone represents five slots. If the burst spans 4 slots × 1 subchannel in the R1 zone and spans 5 slots × 1 subchannel in the R3 zone, where the burst can serve only one user, then the R1 zone can serve 60 users with a data load equal to 4 slots per user, segment A in the R3 zone can serve 10 users with a data load equal to 5 slots per user, and segment BC can serve 20 users with a data load equal to 5 slots per user. Theoretically, the traditional FFR technique (indicated by a red bold line in Figure 4.4) can serve 70 users, and this number can be raised to 90 if segment BC is considered. Consequently, the number of slots in each part of the DL subframe can be deduced, where the number of slots in the R1 zone is equal to 240; in the R3 zone (segment A), 50; and in segment BC, 100. Theoretically, the traditional FFR can provide 290 slots, whereas this number is increased to 390 when segment BC is considered.

4.3 Static Resource Assignment FFR Model

A novel SRA model aims to enhance the performance of traditional FFR base stations in terms of the number of served users, utilized slots, data rate, and spectral efficiency. First, a scenario of resource assignment in the SRA is proposed, and then the SRA algorithm is introduced and explained.

4.3.1 Resource Assignment Scenario in the SRA FFR Model

The cell layout of the proposed SRA model is shown in Figure 4.5, where the radius of the cell center (r) is 635 m and the cell border radius (R) is 1000 m. In order to take the maximum advantage of BC segments, four cases are evaluated: cases 1, 2, 3, and 4, as shown in Figure 4.5. The purpose of using these cases is to find the coverage area that gives the best response for segment BC, where each case represents segment BC with different coverage areas.

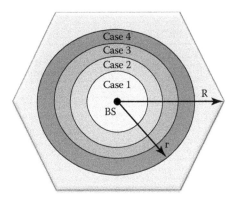

FIGURE 4.5
Proposed SRA cell layout.

The resource assignment rules are defined in Table 4.2, where each case has been defined by two SINR thresholds that enable each case to use two types of burst profiles (BPs). A threshold of 5 dB is used to distinguish between cell center and cell border users.

BP refers to the type of modulation and coding scheme (MCS) that can be used under specific channel conditions (see Section 3.6). The MCS types are selected based on the modulations and SINR thresholds mentioned in [17], where eight types of BPs (or MCSs) are formed, as listed in Table 4.3.

The SINR thresholds in Table 4.3 ensure that the bit error rate is less than 10^{-6} when the convolutional turbo code (CTC) is used.

TABLE 4.2

Resource Assignment Rules per Case

	Segment B		Segment C	
Case No.	SINR Range (dB)	Burst Profile Type (MCS)	SINR Range (dB)	Burst Profile Type (MCS)
Case 1	$SINR \geq 19.9$	BP 8 64 QAM (5/6)	$18 \leq SINR02C2 19.9$	BP 7 64 QAM (3/4)
Case 2	$16.9 \leq SINR < 18$	BP 6 64 QAM (2/3)	$13.8 \leq SINR < 16.9$	BP 5 64 QAM (1/2)
Case 3	$12.7 \leq SINR < 13.8$	BP 4 16 QAM (3/4)	$8.6 \leq SINR < 12.7$	BP 3 16 QAM (1/2)
Case 4	$6.3 \leq SINR < 8.6$	BP 2 QPSK (3/4)	$5 \leq SINR < 6.3$	BP 1 QPSK (1/2)

4.3.2 SRA FFR Algorithm Design

The SINR as a metric is used to assign resources in the DL subframe to a user or mobile station (MS). The resources may be assigned into the R1 zone, the R3 zone, or segment BC, depending on the predefined SINR threshold ($SINR_{TH}$) value. SINR takes into consideration the impact of interference from other base stations (BSs) in the network. However, R1 zone users enjoy eight

TABLE 4.3

SINR Threshold for Different MCS Types

Profile No.	MCS Type Modulation Type	Code Rate Type	Bits per Subcarrier	Minimum SINR Threshold (dB)
BP 1	QPSK	1/2	1	2.9
BP 2		3/4	1.5	6.3
BP 3	16 QAM	1/2	2	8.6
BP 4		3/4	3	12.7
BP 5	64 QAM	1/2	3	13.8
BP 6		2/3	4	16.9
BP 7		3/4	4.5	18
BP 8		5/6	5	19.9

Source: Stiakogiannakis, I. N., et al., *IEEE Antennas and Propagation Magazine*, 54, 214–226, 2012.

types of BPs when a threshold of 5 dB is used to distinguish between inner (R1) and outer (R3) users, as revealed in Table 4.3. The operation sequence of the SRA algorithm is divided into two phases, as shown in Figure 4.6. The first phase shows resource assignment procedures in the DL subframe parts for the proposed SRA FFR algorithm. In contrast, the details of the second

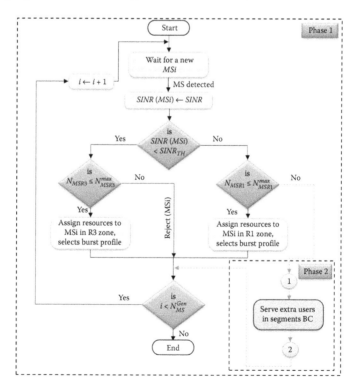

FIGURE 4.6
SRA FFR flowchart to allocate resources in the DL subframe.

phase are shown in Figure 4.7, which describes the resource assignment procedures for the four cases in segment BC. First, the flowchart of phase 1 with its algorithm steps is introduced. The flowchart of phase 2 with its algorithm steps is then presented, as follows:

SRA Algorithm Phase 1 Steps (resource assignment in the DL subframe parts)

Step 1: [Initialization] Defines burst profile set (BPj). Generate random users N_{MS}^{Gen} in the cell of interest, where $N_{MS}^{Gen} = N_{MS_{R1}}^{max} + N_{MS_{R3}}^{max} + N_{MS}^{Extra}$. The generated user is randomly distributed in a two-dimensional plane (MS X-axis and MS Y-axis).

Step 2: BS requests MS to send the SINR value (SINR(MS)) through the REP-REQ control message [18].

Step 3: The MS starts to evaluate the SINR according to its location and the interferer BSs in the network. The MS sends the REP-RSP control message [18] to the BS with the requested information.

Step 4: BS compares the received SINR(MS) value with the $SINR_{TH}$ (5 dB). If the $SINR(MS) < SINR_{TH}$, then **Go to Step 7**; otherwise, **Go to Step 5**.

Step 5: BS assigns resources to MS in the R1 zone and selects the suitable burst profile based on the reported SINR(MS), as long as the number of users in the R1 zone (N_{MSR1}) does not reach its maximum value $N_{MS_{R1}}^{max}$. Then **Go to Step 9**.

Step 6: When the number of users in the R1 zone reaches its maximum value $N_{MS_{R1}}^{max}$, then the extra users will be served by segment BC. Then **Go to** terminal 1 to start the phase 2 algorithm.

Step 7: BS assigns resources to MS in the R3 zone and selects the suitable burst profile based on the reported SINR(MS), as long as the number of users in the R3 zone (N_{MSR3}) does not reach its maximum value $N_{MS_{R3}}^{max}$. Then **Go to Step 9**.

Step 8: When the number of users in the R3 zone reaches its maximum value $N_{MS_{R3}}^{max}$, the MS will be rejected; then **Go to Step 9**.

Step 9: If the number of tested MS(i) ≤ the total number of generated users N_{MS}^{Gen} in the cell coverage area, then **Go to Step 2** to get the next MS (increment *i*). Otherwise, **END**.

SRA Algorithm Phase 1 (resource assignment in the DL subframe parts)

BEGIN
[Initialization] BP ∈ {BP1, ..., BPj}, ∀j = {1, ..., 8} /* *Defines BP types* */
                      /* *Generate n random mobile station in the cell of interested* */

1- $36 \leq \sqrt{MS_X^2 + MS_Y^2} \leq 1000, MS = \{MS1, \ldots MSn\}, \quad n = N_{MS}^{Gen}$

2- **for** i = 1 **to** N_{MS}^{Gen} **do**   /* N_{MS}^{Gen} *is number of generated users* */

3- **Compute** SINR(MSi)

4-   **Zone = 0**

5-   **if** SINR(MSi) < SINR$_{TH}$ **then**

6-   **Zone** = R3   /**define R3 Zone as an operational Zone* */

7-   **else**

8-   **if** $N_{MSR1} \leq N_{MSR1}^{max}$ **then**

9-   **Zone** = R1   /**define R1 Zone as an operational Zone* */

10-   R1 ⟵ MSi   /**Assign user to R1 Zone* */

11-   MSi ⟵ BPj, :BPj≡SINR(MSi) /* *Specify suitable burst profile* */

12-   **Increment** N_{MSR1}

13-   **else**

14-   **Zone** = Seg. BC   /* *Serve extra users by segments BC* */

15-   **Go To** terminal 1   /* *Starts phase 2 of SRA algorithm* */

16-   **endif**

17-   **endif**

18-   **if** (Zone = R3 AND $N_{MSR3} \leq N_{MSR3}^{max}$)   **then**

19-   R3 ⟵ MSi   /* *Assign user to R3 Zone* */

20-   MSi ⟵ BPj, :BPj≡SINR(MSi)   /* *Specify suitable burst profile* */

21-   **Increment** N_{MSR3}

22-   **else**

23-   MSi ∉ R3 *zone*   /* *Reject MSi* */

24-   **endif**

25- **endfor**

END

The second phase of the SRA algorithm aims to utilize segment BC to serve extra users through the four cases, where each case can serve a maximum number of 20 users. The 20 users represent the maximum capacity of segment BC, as mentioned in Section 4.2.2. The number of extra users N_{MS}^{Extra} represents users who are not served by the R1 zone. The operation phases of these cases are shown in the Figure 4.7 flowchart. This flowchart describes the process of case selection and the resource assignment per case, where the decision to select the appropriate segment and suitable BP type is applied.

The process flow in the Figure 4.7 flowchart starts from connector terminal 1 and ends by connector terminal 2, as mentioned in Figure 4.6. The algorithm steps of the second phase are as follows.

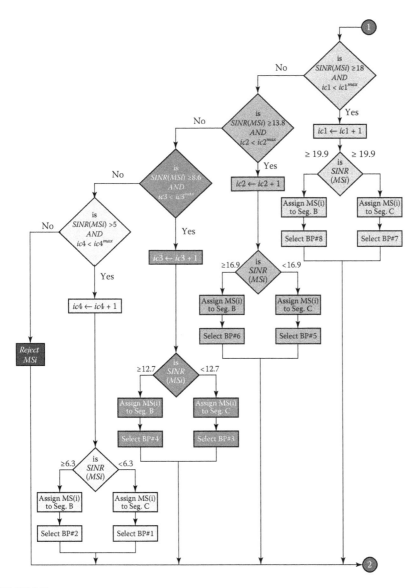

FIGURE 4.7
Four cases of resource assignment flowchart.

SRA Algorithm Phase 2 Steps (resource assignment per case)

Step 1: [Initialization] Set the ($ic1$, $ic2$, $ic3$, and $ic4$) to zero, which represent the user counters for cases 1, 2, 3, and 4, respectively. These counters are used to reject users in each case when a case is fully filled with users. The BS holds the SINR(MSi) of the current MS.

Step 2: [Case 1] When ($18\,dB \le$ SINR(MSi)), assign resources for MS using either segment B or C with BP 8 or 7, respectively, and then **Go to Step 7**. Otherwise, **Go to** next case. The activity of case 1 is disabled when the number of users exceeds $ic1^{max}$ (20).

Step 3: [Case 2] When ($13.8\,dB \le$ SINR(MSi) $<18\,dB$), assign resources for MS using either segment B or C with BP 6 or 5, respectively, **and** then **Go to Step 7**. Otherwise, **Go to** next case. The activity of case 2 is disabled when the number of users exceeds $ic2^{max}$ (20).

Step 4: [Case 3] When ($8.6\,dB \le$ SINR(MSi) $< 13.8\,dB$), assign resources for MS using either segment B or C with BP 4 or 3, respectively, and then **Go to Step 7**. Otherwise, **Go to** next case. The activity of case 3 is disabled when the number of users exceeds $ic3^{max}$ (20).

Step 5: [Case 4] When ($5\,dB \le$ SINR(MSi) $< 8.6\,dB$), assign resources for MS using either segment B or C with BP 2 or 1, respectively, and then **Go to Step 7**. Otherwise, **Go to Step 6**. The activity of case 4 is disabled when the number of users exceeds $ic4^{max}$ (20).

Step 6: This step rejects a user when the number of users in the current case reaches its maximum value or when the user SINR did not match the predefined SINR threshold values of the four cases. Then **Go to Step 7**.

Step 7: Go to terminal 2 (return to **Step 9** in phase 1 of the SRA algorithm).

SRA Algorithm Phase 2 (resource assignment per case)

BEGIN
[Initialization] Set ($ic1 = 0$, $ic2 = 0$, $ic3 = 0$, $ic4 = 0$), receives data from terminal 1

1-   if (SINR(MSi) $\ge$ 18 **AND** $ic1 < 20$)   **then** /* *Case 1 Conditions* */
2-   **Increment** $ic1$
3-   if SINR(MSi) $\ge$ 19.9
4-   Segment B $\longleftarrow$ MSi   /**Assign users to segment B*/
5-   MSi$\longleftarrow$BP(8)   /* *Selects suitable BP* */

6-   **else**
7-   Segment C ⟵ MSi /*Assign users to segment C*/
8-   MSi ⟵ BP(7)   /*Selects suitable BP*/
9-   **endif**
10- **else**
11-   **if** (SINR(MSi) ≥ 13.8 **AND** $ic2 < 20$) **then**
                                    /* Case 2 Conditions */
12-   **Increment** $ic2$
13-   **if** SINR(MSi) ≥ 16.9
14-   Segment B ⟵ MSi
15-   MSi ⟵ BP(6)
16-   **else**
17-   Segment C ⟵ MSi
18-   MSi ⟵ BP(5)
19-   **endif**
20-   **else**
21-   **if** (SINR(MSi) ≥ 8.6 **AND** $ic3 < 20$) **then**
                                    /* Case 3 Conditions */
22-   **Increment** $ic3$
23-   **if** SINR(MSi) ≥ 12.7
24-   Segment B ⟵ MSi
25-   MSi ⟵ BP(4)
26-   **else**
27-   Segment C ⟵ MSi
28-   MSi ⟵ BP(3)
29-   **endif**
30-   **else**
31-   **if** (SINR(MSi) ≥ 5 **AND** $ic4 < 20$) **then**
                                    /* Case 4 Conditions */
32-   **Increment** $ic4$
33-   **if** SINR(MSi) ≥ 6.3
34-   Segment B ⟵ MSi
35-   MSi ⟵ BP(2)
36-   **else**
37-   Segment C ⟵ MSi

```
38-                              MSi ⟵ BP(1)
39-                    endif
40-            else  MSi ∉ Case,      : Case ∈ {Case1, …, Case4}
                                             /* Reject Ms) */
41-            endif
42-      endif
43-   endif
44- endif
```

END   **Go to** Terminal 2

The decision to choose any of these cases to be used by segment BC depends on the response of each case. The case that greatly enhances base station performance will be chosen as a preferable configuration for the proposed SRA FFR. The configuration in the SRA FFR is specified in the design time. Once a high-performance case is recognized as a result of the initial test, the network administrator uses this case as a desirable configuration for a long period of time. This type of configuration is termed static FFR (see Section 2.5), where the system parameters (configuration) are set earlier and remain fixed during a specific period of time. The proposed SRA FFR does not take into account the variation in population density when assigning resources to segment BC. The availability of active users is subject to continuous change as they move from one place to another. In order to consider the variation in population density, a dynamic FFR model is proposed, as in the following section.

4.4 Dynamic Resource Assignment FFR Model

Although the DRA model addresses the variation in population density, it enhances the performance of the traditional FFR (802.16e) base station, as well as the number of users served, resource utilization, data rate, and spectral efficiency. The system parameters and the DL subframe structure of DRA are similar to those in Table 4.1 and Figure 4.4, respectively. The load per user in the DRA model is the same as in the SRA model. The DRA FFR can serve 60 users in the R1 zone with a load of four slots per user, 10 users in the R3 zone with a load of five slots per user, and 20 users in segment BC with a load of five slots per user, as mentioned in Section 4.2.2.

The DRA FFR model utilizes segment BC to serve extra users in the cell center area. The key differences in the simulation scenario between DRA and SRA FFRs are

1. DRA FFR solves the problem of variation in population density, which can vary according to the served area type, such as markets, schools, hospitals, and government departments.

2. Segment BC coverage area is equal to the entire cell center area and uses eight types of burst profiles, unlike SRA FFR, where segment BC serves only a slice of the cell center area with two types of burst profiles. In SRA FFR, segment BC serves a slice of the cell center area through one of the cases, as mentioned in Section 4.3.1.

In the following subsections, the resource assignment scenario is suggested and the DRA algorithm design is then presented.

4.4.1 Resource Assignment Scenario in the DRA FFR Model

In the DRA FFR model, the cell center area is partitioned into four layers: A, B, C, and D, and each layer covers different parts of the cell center area. The proposed DRA FFR cell layout is shown in Figure 4.8. The reason behind partitioning the cell center area into four layers is to compute the number of users in each layer, whereby the layer with the highest population density (winner layer) will be served by segment BC. This process is repeated automatically frame by frame in order to continually update the information of the most crowded layer. In other words, the base station will change the coverage area of segment BC according to the crowded layer. This type of configuration is called dynamic FFR (see Section 2.5). In dynamic FFR, the system parameters are dynamically changed frame by frame according to the cell load. Accordingly, resource utilization in the base station will be more efficient, since users are served based on distribution density.

The rules for assigning resources into the appropriate layer in DRA FFR are based on the SINR thresholds set out in Table 4.4. There are four layers and eight types of BPs (MCSs), where each layer is defined by two SINR thresholds (upper and lower). In addition, each layer can use two types of BPs. The SINR thresholds and MCSs listed in Table 4.4 are specified according to the MCSs and SINR thresholds mentioned in Table 4.3 [17]. However, layers near the base station can use high MCS types, whereas layers far away from the base station use low MCS types. These arrangements for MCS usage are related to the phenomenon of path loss [19]. Users near the base station (such as layer A) have low path loss and, at the same time, have large path loss in the direction of neighboring base stations that operate as a source of interference. As a result, users near the base station can enjoy high SINR values, and therefore can use high modulation order. Nevertheless, users far away from the base station (such as layer D) suffer from large path loss or low SINR values, which forces the base station to use low modulation order, as revealed in Table 4.4.

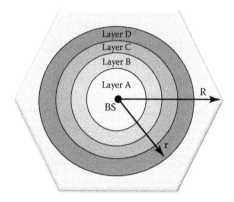

FIGURE 4.8
Proposed DRA FFR cell layout.

TABLE 4.4

Resource Assignment Rules per Layer

Layer Name	Segment B			Segment C		
	SINR Threshold (dB)	Burst Profile Type (MCS)		SINR Threshold (dB)	Burst Profile Type (MCS)	
A	$SINR \geq 19.9$	BP 8	64 QAM (5/6)	$18 \leq SINR < 19.9$	BP 7	64 QAM (3/4)
B	$16.9 \leq SINR < 18$	BP 6	64 QAM (2/3)	$13.8 \leq SINR < 16.9$	BP 5	64 QAM (1/2)
C	$12.7 \leq SINR < 13.8$	BP 4	16 QAM (3/4)	$8.6 \leq SINR < 12.7$	BP 3	16 QAM (1/2)
D	$6.3 \leq SINR < 8.6$	BP 2	QPSK (3/4)	$5 \leq SINR < 6.3$	BP 1	QPSK (1/2)

4.4.2 DRA FFR Algorithm Design

User assignment into the R1 zone, R3 zone, or segment BC, is based on a predefined SINR threshold. Five decibels is used to distinguish between cell center users (R1 zone and segment BC) and cell border users (R3 zone). However, using a threshold of 5 dB enables the R1 zone and segment BC users to use eight types of BPs, as mentioned in Table 4.3. Segment BC serves extra users in the cell center area, and these users are classified into four layers (A, B, C, and D) based on their distance from the base station, where the DRA algorithm computes the distance of these users to the base station at each trial (or attempt).

The operation sequences of the DRA FFR are divided into three stages. These three stages are illustrated in Figure 4.9. The first and second stages involve the selection of a high-population layer and resource assignment in the DL subframe parts, respectively. The third stage involves resource assignment into segment BC through the dynamic layers (A, B, C, and D). The details of the third stage are illustrated in Figure 4.10.

The tasks of the first two stages in the flowchart of Figure 4.9 are as follows: the first stage illustrates the determination of the crowded layer preceded by the process of isolating R1 zone users in the cell center area, since both the R1 zone and segment BC serve users in the same area (cell center area). Practically, each layer is defined by upper and lower boundaries. These boundaries can be observed in Figure 4.3, where each SINR value is related to a particular distance, since the SINR value varies according to the distance at which the SINR is measured. The second stage shows the resource assignment procedures into the R1 and R3 zones and segment BC. However, the algorithm steps of the first and second stages of DRA FFR are as follows:

DRA Steps of Stage 1 (determining the crowded layer)

Step 1: [Initialization] Defines burst profile set (BPj). Set the user counter (N_{R1}) of the R1 zone to zero. Set N_{MS}^{ABCD} to zero. Set ($N_{MS}^{LA}, N_{MS}^{LB}, N_{MS}^{LC}, N_{MS}^{LD}$) to zero, which count the number of users in layers A, B, C, and D, respectively. Set the user index flag MSF(j) to null, which defines each user to a specific layer type.

Step 2: Generate random users N_{MS}^{Gen} in the target cell, where $N_{MS}^{Gen} = N_{MS_{R1}}^{max} + N_{MS_{R3}}^{max} + N_{MS}^{Extra}$. The generated user is randomly distributed in a two-dimensional plane (MS X-axis and MS Y-axis).

Step 3: If the location of a user is in the cell center area, then increment N_{R1} and **Get** the next user as long as the number of users in the R1 zone (N_{R1}) has not reached its maximum value $N_{MS_{R1}}^{max}$. Otherwise, the user does not belong to the R1 zone and so **Go to Step 11**.

Step 4: When the number of users in the R1 zone reaches its maximum value, terminate the loop and **Go to Step 5**.

Step 5: When the user location is within the boundaries of layer A, increment the number of users in layer A (N_{MS}^{LA}) and link this user with a flag layer type of A, and then **Go to Step 10**. Otherwise, **Go to Step 6**.

Step 6: When the user location is within the boundaries of layer B, increment the number of users in layer B (N_{MS}^{LB}) and link this user with a flag layer type of B, and then **Go to Step 10**. Otherwise, **Go to Step 7**.

Step 7: When the user location is within the boundaries of layer C, increment the number of users in layer C (N_{MS}^{LC}) and link this user with a flag layer type of C, and then **Go to Step 10**. Otherwise, **Go to Step 8**.

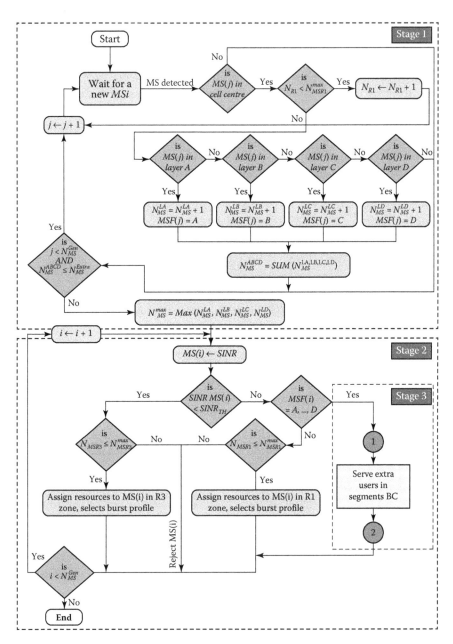

FIGURE 4.9
DRA FFR flowchart to allocate resources in the DL subframe.

Step 8: When the user location is within the boundaries of layer D, increment the number of users in layer D (N_{MS}^{LD}) and link this user with a flag layer type of D, and then **Go to Step 10**. Otherwise, **Go to Step 9**.

Step 9: If the user location is out of the coverage area of all layers, **Go to Step 11**.

Step 10: Save the number of users in all of the layers in N_{MS}^{ABCD}, and then **Go to Step 11**.

Step 11: When the number of tested MS(i) is less than the total number of generated users N_{MS}^{Gen} and the sum of users in the four layers ($N_{MS}^{A,B,C,D}$) does not exceed the number of extra users, **Get** the next user. Otherwise, terminate the loop and make a decision to identify the most crowded layer; save the name of the crowded layer in (N_{MS}^{max}). Then **Go to Step 12**.

Step 12: The process continues to **Stage 2**.

DRA Algorithm of Stage 1 (determining the crowded layer)

BEGIN
[Initialization] BP ∈ {BP1, ..., BPj}, ∀j={1,...,8}   /* *defines BP types* */
   /* *Generate n random mobile station in the target cell* */

1- $36 \leq \sqrt{MS_X^2 + MS_Y^2} \leq 1000$, MS = {MS1,...,MSn}, $n = N_{MS}^{Gen}$

2- **for** j = 1 **to** N_{MS}^{Gen} **do**   /* *determine cell center members* */

3- **if** $36 \leq \sqrt{MS_{xj}^2 + MS_{yj}^2} < 635$ **then**

4- **if** $N_{R1} \leq N_{MS_{R1}}^{max}$ **then**

5-   **Increment** N_{R1}

6- **else**

7- **if** ($N_{R1} = N_{MS_{R1}}^{max}$ **AND** $N_{MS}^{ABCD} \leq N_{MS}^{Extra}$) **then**
               /* *determine Seg. BC members* */

8-   **if** $36 \leq \sqrt{MS_{xj}^2 + MS_{yj}^2} < 290$ /* *Layer A Condition* */

9-   **Increment** N_{MS}^{LA}

10-   A⟵MSF(j)   /* *sets flag type to A* */

11-   **else**

12-   **if** $290 \leq \sqrt{MS_{xj}^2 + MS_{yj}^2} < 380$ /* *Layer B Condition* */

13-   **Increment** N_{MS}^{LB}

14-   B ⟵ MSF(j)   /* sets flag type to B */

15-   **else**

16-   **if** $380 \leq \sqrt{MS_{xj}^2 + MS_{yj}^2} < 520$ /* Layer C Condition */

17-   **Increment** N_{MS}^{LC}

18-   C ⟵ MSF(j)   /* sets flag type to C */

19-   **else**

20-   **if** $520 \leq \sqrt{MS_{xj}^2 + MS_{yj}^2} \leq 635$ /* Layer D Condition */

21-   **Increment** N_{MS}^{LD}

22-   D ⟵ MSF(j)   /* sets flag type to D */

23-   **else**

24- $MS(j) \notin Layer, : Layer \in \{Layer\,A, ..., Layer\,D\}$

                                                /*reject MS(j)*/

25-   **endif**

26-   **endif**

27-   **endif**

28-   **endif**

29- $N_{MS}^{ABCD} = N_{MS}^{LA} + N_{MS}^{LB} + N_{MS}^{LC} + N_{MS}^{LD}$

30-   **endif**

31- **enfif**

32- **else**

33- $MS(j) \notin$ Cell center area   /* reject MS(j)*/

34- **endif**

35- **endfor**

36- $N_{MS}^{max} = Max\left(N_{MS}^{LA}, N_{MS}^{LD}, N_{MS}^{LC}, N_{MS}^{LD}\right)$ /* determine the crowded layer */

END

The second stage of the DRA algorithm is employed to make the decision to house the user in the R1 zone, R3 zone, or segment BC, as follows.

DRA Steps of Stage 2 (resource assignment into R1, R3, and segment BC)

Step 1: The BS requests the MS to send the SINR value (SINR(MS)) through the REP-REQ control message [18].

Step 2: The MS starts to evaluate the SINR according to its location and the interferer BSs in the network. The MS sends a REP-RSP message [18] to the BS holding the requested information.

Step 3: The BS compares the received SINR(MS) value with the $SINR_{TH}$. If $SINR(MS) < SINR_{TH}$, then **Go to Step 6**; otherwise, **Go to Step 4**.

Step 4: If the current MS is flagged with one of the layers types (A, B, C, or D), then the BS considers this MS as one of the segment BC members; next, **Go To** terminal 1 to start **Stage 3**. Otherwise, the BS assigns resources to MS in the R1 zone and selects a suitable burst profile based on the reported SINR(MS), as long as the number of users in the R1 zone has not reached its maximum value $N_{MS_{R1}}^{max}$. Then **Go to Step 8**.

Step 5: If the number of users in the R1 zone reaches its maximum value $N_{MS_{R1}}^{max}$, then the MS will be rejected. **Go to Step 8**.

Step 6: The BS assigns resources to MS in the R3 zone and selects a suitable burst profile based on the reported SINR(MS), as long as the number of users in the R3 zone does not reach its maximum value $N_{MS_{R3}}^{max}$. Then **Go to Step 8**.

Step 7: If the number of users in the R3 zone reaches its maximum value $N_{MS_{R3}}^{max}$, the MS will be rejected. Then **Go to Step 8**.

Step 8: If the number of tested MS(i) is less than the total number of generated users N_{MS}^{Gen} in the cell coverage area, then **Go to Step 1** to get the next MS (increment i). Otherwise, **END**.

DRA Algorithm of Stage 2 (resource assignment into R1, R3, and segment BC)

BEGIN

1- **for** i = 1 **to** N_{MS}^{Gen} **do**   /* N_{MS}^{Gen} *is number of generated users* */

2-   **Compute** SINR(MSi)

3-   **Zone** = 0

4-   **if** SINR(MSi) < 5 dB **then**

5-   **Zone** = R3   /*define R3 Zone as an operational Zone */

6-   **else**

7-   **if** MSF(i) = A or B or C or D **then**

8-   **Zone** = Seg. BC   /*define Segment BC as an operational area */

9- **Go To** terminal 1 /* *Starts stage 3 of DRA algorithm* */

10- **else**

11- **if** $N_{\mathrm{MSR1}} \leq N_{\mathrm{MSR1}}^{\max}$ **then**

12- **Zone** = R1 /*define R1 Zone as an operational Zone */

13- R1 $\longleftarrow$ MS(i)

1- MS(i) $\longleftarrow$ BPj, : BPj $\equiv$ SINRMS(i) /* *Specify suitable burst profile*/

14- **Increment** N_{MSR1}

15- **else**

16- MS(i) $\notin$ R1 *zone* /* *Reject MS(i)* */

17- **endif**

18- **endif**

19- **endif**

20- **if** (Zone = R3 **AND** $N_{\mathrm{MSR3}} \leq N_{\mathrm{MSR3}}^{\max}$) **then**

21- R3 $\longleftarrow$ MS(i)

2- MS(i) $\longleftarrow$ BPj, : BPj $\equiv$ SINR(MSi) /* *Specify suitable burst profile*/

22- **Increment** N_{MSR3}

23- **else**

24- MS(i) $\notin$ R3 *zone* /* *Reject MS(i)* */

25- **endif**

26- **endfor**

END

The third stage of the DRA FFR algorithm offers its services to the crowded layer through the utilization of segment BC. The number of extra users ($N_{\mathrm{MS}}^{\mathrm{Extra}}$) represents the users who are not served by the R1 zone. The operation sequences of resource assignment to these layers and selecting a suitable burst profile type are illustrated in Figure 4.10, where the resources are allocated to the most crowded layer. The process flow of stage 3 starts from connector terminal 1 and ends by connector terminal 2, as illustrated in Figure 4.9. The algorithm and steps of stage 3 are as follows:

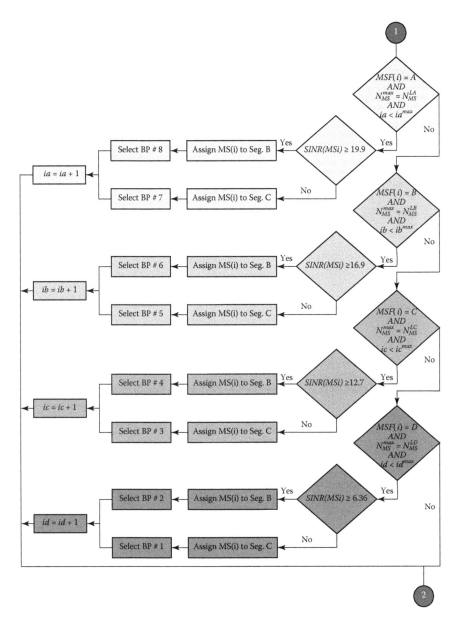

FIGURE 4.10
Crowded layer resource allocation flowchart.

DRA Algorithm Stage 3 (resources allocation per layer)

BEGIN

1- **Assume** $ia = 0$, $ib = 0$, $ic = 0$, $id = 0$, Receives data from terminal 1

2- **if** $\left(\mathrm{MSF}(i) = A \textbf{ AND } N_{\mathrm{MS}}^{\max} = N_{\mathrm{MS}}^{\mathrm{LA}} \textbf{ AND } ia < ia^{\max} \right)$ **then**
       /*Layer A Condition*/

3-   **if** SINR(MSi) $\geq$ 19.9   **then**

4-   Segment B $\longleftarrow$ MS(i)   /* Assign resources
                                                                                        to MS(i) in Seg. B */

5-   MS(i) $\longleftarrow$ BP(8)   /* Selects suitable burst
                                                                                              profile */

6-   **else**

7-   Segment C $\longleftarrow$ MS(i)

8-   MS(i) $\longleftarrow$ BP(7)

9-   **endif**

10-   **Increment** ia

11- **else**

12-   **if** $\left(\mathrm{MSF}(i) = B \textbf{ AND } N_{\mathrm{MS}}^{\max} = N_{\mathrm{MS}}^{\mathrm{LB}} \textbf{ AND } ib < ib^{\max} \right)$ **then**
                                                                       /*Layer B Condition*/

13-   **if** SINR(MSi) $\geq$ 16.9   **then**

14-   Segment B $\longleftarrow$ MS(i)

15-   MS(i) $\longleftarrow$ BP(6)

16-   **else**

17-   Segment C $\longleftarrow$ MS(i)

18-   MS(i) $\longleftarrow$ BP(5)

19-   **endif**

20-   **Increment** ib

21- **else**

22- **if** $\left(\mathrm{MSF}(i) = C \textbf{ AND } N_{\mathrm{MS}}^{\max} = N_{\mathrm{MS}}^{\mathrm{LC}} \textbf{ AND } ic < ic^{\max} \right)$ **then**
                                                                       /* Layer C condition */

23-   **if** SINR(MSi) $\geq$ 12.7   **then**

24-   Segment B $\longleftarrow$ MS(i)

25-   MSi $\longleftarrow$ BP(4)

26-   **else**

27- Segment C ⟵ MS(i)

28- MSi ⟵ BP(3)

29- **endif**

30- **Increment** *ic*

31- **else**

32- **if** $\left(\text{MSF}(i) = D \text{ AND } N_{\text{MS}}^{\text{max}} = N_{\text{MS}}^{\text{LD}} \text{ AND } id < id^{\text{max}} \right)$ **then**
/* *Layer D Condition* */

33- **if** SINR(MSi) ≥ 6.3   **then**

34- Segment B ⟵ MS(i)

35- MS(i) ⟵ BP(2)

36- **else**

37- Segment C ⟵ MS(i)

38- MS(i) ⟵ BP(1)

39- **endif**

40- **Increment** *id*

41- **else**

42- $\text{MS}(i) \notin \text{Layer, : Layer} \in \{\text{Layer A}, ..., \text{Layer D}\}$,   /**reject* *MS(i)* */

43- **endif**

44- **endif**

45- **endif**

46- **endif**

END   **Go to** Terminal 2

DRA Steps of Stage 3 (resource allocation per layer)

Step 1: [Initialization] Set $(ia = 0, ib = 0, ic = 0, id = 0)$, which are used to count the number of users in layers A, B, C, and D, respectively, and signify the user's rejection when the layer is filled with users. The base station holds the identity of the most crowded layer ($N_{\text{MS}}^{\text{max}}$), MS flag index (MSF), and SINR of the current MS.

Step 2: When both the identity of the crowded layer and the user flag index are equal to $N_{\text{MS}}^{\text{LA}}$ and A, respectively, then the resources are assigned to MS in layer A. Based on the MS signal quality (SINR), the resources are assigned either to segment B with BP 8 or to segment C with BP 7, as long as the number of users in layer A does not exceed the maximum value ia^{max}. Then **Go to Step 5**.

Step 3: When the number of users in layer A reaches its maximum value, the user will be rejected. Then **Go to Step 4** (next layer).

Step 4: Repeat **Steps 2** and **3** for layers B, C, and D; then **Go to Step 5**.

Step 5: Go to terminal 2 (return back to **Step 8 in Stage 2** of the DRA algorithm).

Finally, the SRA and DRA algorithms can be implemented in the media access control (MAC) layer of the base station (see Section 3.7), and the outcomes of these algorithms are achieved without additional devices, or changes to the network infrastructure, or increasing the complexity of the system. However, in order to avoid the complexity of algorithms, the details of resource assignment into the R1 and R3 zones are not included in the SRA and DRA algorithms. The detailed procedures of resource assignment into these zones are similar to those for segment BC. For instance, if the intended user belongs to the R3 zone, the base station will then assign a suitable modulation type based on the user SINR value and will allocate resources to this user in the R3 zone. The user will be rejected if the reported SINR does not coincide with the terms of the allocation of resources in the R3 zone or if there are not enough free slots in this zone. Moreover, and from a design point of view, the specifications of cases and layers in terms of area of service are the same. However, the terms *case* and *layer* are used to distinguish between static and dynamic configurations, respectively.

4.5 Analyzing and Modeling Performance Metrics

In this section, selective metrics are analyzed and modeled, in order to be used as criteria to evaluate the performance of the proposed algorithms. These metrics are SINR, data rate, subcarrier efficiency, spectral efficiency, frame capacity, and channel capacity. The performance can be evaluated by considering either peak or average values. The peak value represents the best performance of the base station, which is not available for all users and at any time. In contrast, and from the perspective of the users, the average performance is what users will typically experience with wireless services. Therefore, the peak and average values of the aforementioned metrics are considered in this book.

4.5.1 Signal-to-Interference-Plus-Noise Ratio Modeling

The SINR represents the ratio of the received power for a particular user location in the cell. The user receives two kinds of power: desired power, which is used to enable the user to communicate with the base station, and undesired power, which represents the interference power transmitted from all base stations using the same frequency band in the same time slots plus noise power. In order to calculate the SINR, three components should be computed: received desired power, received undesired power, and noise power. In general, received power can be calculated as in Equation 4.3 [20].

$$Pr = Pt + Gt + Gr - PL \quad \text{dB} \tag{4.3}$$

where:
  Pt is the transmitted signal power.
  Gt and Gr denote the transmitter and receiver antenna gain, respectively.
  PL is the path loss.

IEEE 802.16e technology is designed to operate efficiently in NLOS connections to support mobile application. From the power point of view, mobile users experience different power levels when they move, since the power level changes when the distance between the base station and the user is changed. This distance-dependent power loss is called path loss [19], where all users in the cell are subject to path loss. The phenomenon of path loss helps network designers to reuse the available spectrum when deploying base stations in cellular networks because the interference power will decay after a certain distance, which allows reuse of the same spectrum in other cells.

The WiMAX standard defines the extended Hata to simulate the channel propagation model. The extended Hata was modified by the European Cooperative for Scientific and Technical (COST) research group and named the COST-231 Hata model, which is given in Equation 4.4 [3].

$$PL = 46.3 + 33.9 \log_{10}(f) - 13.82 \log_{10}(h_{BS}) + $$
$$(44.9 - 6.55 \log_{10}(h_{BS})) \times \log_{10}(d) - A(h_{MS}) + C_F + X \quad \text{dB} \tag{4.4}$$

where:
  f is the carrier frequency in megahertz.
  h_{BS} denotes the base station antenna height in meters.
  d is the distance between the sender and the receiver in kilometers.
  C_F is the environment correction factor, where it equals 3 and 0 dB for urban and suburban areas, respectively [21].

The WiMAX Forum recommends adding a 10 dB fade margin for the path loss to account for shadowing (X) [21]. $A(h_{MS})$ is the mobile antenna correction factor and can be calculated as in Equation 4.5 [3].

$$A(hm) = (1.11 \times \log_{10}(f) - 0.7) \times h_{MS} - (1.56 \times \log_{10}(f) - 0.8) \tag{4.5}$$

where h_{MS} is the mobile station antenna height in meters. The values of the propagation parameters are given in Equations 4.4 and 4.5 and listed in Table 4.5 [3].

If we compensate the values in Table 4.5 in Equations 4.4 and 4.5, the *PL* can be rewritten as in Equation 4.6, and if we compensate the value of *PL* in Equation 4.3, then the received power in Equation 4.3 can be modified as in Equation 4.7.

$$PL = 52.1971 + 35.0412 \times \log_{10}(d) \, \text{dB} \qquad (4.6)$$

$$Pr = Pt + Gt + Gr - \left[52.1971 + 35.0412 \times \log_{10}(d) \right] \, \text{dB} \qquad (4.7)$$

In order to find the desired and undesired received powers, the distance (*d*) in Equation 4.7 must be calculated in two directions. The first direction is the distance between the user location and the serving base station, and the second direction is the distance between the user location and the interferer base stations. Therefore, the locations of the base stations in the grid need to be specified. Figure 4.11 shows the necessary calculations for cell allocation in the grid. The hexagon forms six equal sides, and the points of convergence of two sides form an angle of 120°. There are six 120° angles and six equilateral triangles in each hexagon. The following steps illustrate the cell allocation procedures in the grid:

Step 1: In Figure 4.11a the altitude (height) *H* can be obtained by $\cos(30) = (H/R)$, where *H* equals $\dfrac{\sqrt{3}R}{2}$.

Step 2: Figure 4.11b illustrates an example of finding the location of the base stations in the first and second tier in the grid. Exploring Δabc to find the location of base station 1, the angle $\varnothing$ equals 30°

TABLE 4.5

Propagation Model Parameters

Item	Value
Path loss model	COST HATA 231, urban
Thermal noise power (*No*)	−174 dBm/Hz
Shadowing (*X*)	10 dB
Base station power (*Pt*)	43 dBm
Base station antenna gain (*Gt*)	15 dBi
Base station antenna height (h_{BS})	32 m
Mobile antenna height (h_{MS})	1.5 m
Mobile antenna gain (*Gr*)	0 dBi

Source: Forum, WiMAX™ system evaluation methodology, version 2.1, 2008, p. 209.

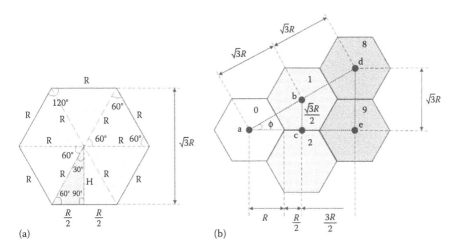

FIGURE 4.11
Hexagon system. (a) One cell dimension. (b) Cell allocation example.

$$\left(\tan^{-1}\left(\frac{\sqrt{3}R/2}{3R/2}\right)\right), \quad \text{and} \quad \tan\varnothing = \frac{bc}{ac}, \quad \text{which gives the height of}$$

$bc = \dfrac{\sqrt{3}R}{2}$ in the Y-axis. Also, the location of base station 1 in the

X-axis is at $3R/2$, where $\cos\varnothing = \dfrac{ac}{ab}$, which results in $ac = \dfrac{3R}{2}$.

Step 3: Base station 1 is located at a Y coordinate point of $\sqrt{3}R/2$ and an X coordinate point of $3R/2$, and the distance (hypotenuse, or *ab*) between base stations 0 and 1 is equal to $\sqrt{3}R$, where $\sin\varnothing = \dfrac{bc}{ab}$, which leads to $ab = \sqrt{3}R$.

Step 4: If R equals 1000 m, then base station 1 is located at the X, Y coordinate points of (1500, 866), and the distance to the center of base station 0 is 1732 m, with $\varnothing$ equal to 30°.

Step 5: In order to find the location of base station 8, the former calculations are repeated by considering Δade;

Step 6: Base station 8 is located at the X, Y coordinate points of (3000, 1732), and the distance (*ad*) between base stations 0 and 8 is 3464 m, with $\varnothing$ equal to 30°. Consequently, all the base station locations are computed as in Figure 4.12.

The cell in the center of the grid (0, 0) is the target cell where all the calculations are obtained. Since the locations of all base stations in the grid are specified, the distance (*d*) from a user location to any base station in the grid can be obtained using Equation 4.8.

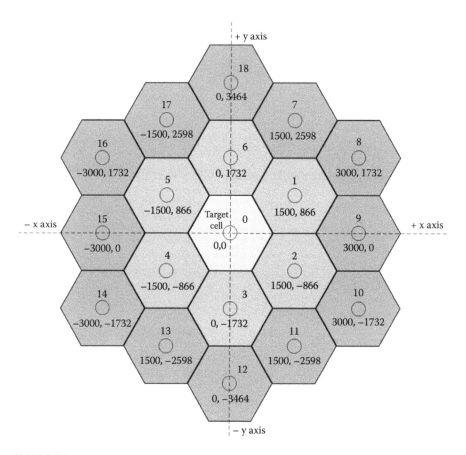

FIGURE 4.12
Network layout in the Cartesian coordinate system.

$$d = \sqrt{(\Delta X)^2 + (\Delta Y)^2} \equiv \sqrt{(MS_x - BS_x)^2 + (MS_y - BS_y)^2} \text{ km} \qquad (4.8)$$

Equation 4.8 is used to find the distance between two points in the XY-plane, where MS_{xy} and BS_{xy} represent the user and base station locations in the XY-plane, respectively. Finally, after finding the required distances to compute the desired and undesired powers, the SINR can be calculated as in Equation 4.9 [1]:

$$SINR = \frac{Pt_i \, Gt_i \, PL_i}{\sum_{j=1}^{I} Pt_j \, Gt_j \, PL_j + No} \qquad (4.9)$$

where:
Pt_i and Gt_i are the transmitted power of BS_i and the channel gain between user i and BS_i, respectively.

The transmitted interference power from other base stations is denoted as Pt_j and Gt_j, where $j \neq i$ and l indicates the number of interfering base stations in the grid.

PL is the path loss.

No is the thermal noise, which can be determined by KTB, where K is the Boltzmann constant (1.38×10^{-23} J/K), T is the absolute Kelvin temperature (290 K), and B is the system bandwidth in hertz; for 1 Hz, the power spectrum density is -174 dBm/Hz.

4.5.2 WiMAX Data Rate Modeling and Calculation

In network deployment, the data rate that a base station can transmit depends on the distribution and the types of users in the cell coverage area, as well as their modulations. In addition, the data rate can vary according to the system parameters, such as system bandwidth, frame duration, OFDM symbol duration, and FFT size. First, the WiMAX 802.16e DL subframe data rate calculation is illustrated (without FFR), and the derivation of the FFR data rate formula can then be shown.

The 802.16e DL subframe data rate can be represented in two ways: physical (PHY) data rate and MAC data rate. The PHY data rate is sometimes called the instantaneous data rate, since it is calculated for one OFDM symbol time, while the MAC data rate is calculated based on frame duration time [15]. First, the OFDM symbol time duration is calculated, which is then followed by calculating the PHY and MAC data rates.

The OFDM symbol time (Ts) is a key element to calculate the data rate, as it combines useful symbol time (Tb) and guard time (Tg), where Tg is redundant data added to the useful data to overcome the multipath effect [21]. The ratio of (tg/tb) is denoted as G and has an impact on the data rate, since a high G ratio reduces the useful data rate [15]. The Ts can be obtained as follows [15]:

$$
\left.
\begin{aligned}
& Ts = Tb + Tg \\
& Ts = Tb + (G \times Tb), \text{ and } Tb = 1/\Delta f \\
& Ts = \frac{1}{\Delta F} \times (1 + G) \\
& \text{Where } \Delta F = \frac{Fs}{N_{FFT}}, \text{ and } Fs = \text{floor}\left(n \times \frac{BW}{8000}\right) \times 8000 \\
& Ts = \frac{1}{n \times \dfrac{BW}{N_{FFT}}} \times (1 + G)\,\mu s
\end{aligned}
\right\}
\tag{4.10}
$$

where:

BW is the system bandwidth.

N_{FFT} represents the total number of subcarriers.

Δf is the frequency spacing.

F_S is the sampling frequency approximated down to an integer product of 8 kHz by using the floor function.

G is the cyclic prefix.

n is the sampling factor, where $G = 1/8$ and $n = 28/25$, as recommended in the standard for a system bandwidth of 10 MHz [18].

The PHY and MAC data rates of the DL subframe can be obtained by using the following equations [22,23]:

$$Dr_{PHY} = \frac{Kr_{OFDM} \times B}{T_S} \text{ bps} \tag{4.11}$$

$$Dr_{MAC} = \frac{Kr_{slot} \times \gamma \times B \times \left(N_{OFDM}^{DL} - N_{OFDM}^{OH}\right)/2}{T_f} \text{ bps} \tag{4.12}$$

Herein, Kr_{OFDM} and Kr_{slot} are the number of data subcarriers per OFDM symbol and slot, respectively. γ represents the number of slots in two successive OFDM symbols (required for the PUSC mode). N_{OFDM}^{DL} indicates the number of OFDM symbols in the DL subframe, and N_{OFDM}^{OH} indicates the number of OFDM symbols allocated for control messages (overhead). T_f is the WiMAX frame duration time, which equals $(T_S \times S_{frame})$, where Ts is as in Equation 4.10 and S_{frame} is the number of OFDM symbols in the WiMAX frame. The results of the numerator in Equations 4.11 and 4.12 mainly depend on the channel condition and the number of subcarriers (design issues), whereas the results of the denominator depend on the specifications of the system parameters. Thus, system parameters greatly affect the aggregate data rate. Finally, the number of data bits per subcarrier B can be obtained by using Equation 4.13 [22].

$$B = Cr \log_2 (Q) \text{ bits/subcarrier} \tag{4.13}$$

where:

Q is the number of points in the constellation for a particular modulation type

Cr is the code rate

The output of Equations 4.11 and 4.12 depends on the modulation and code rate type used in a specific link. Table 4.6 shows the DL subframe slot capacity, number of bits per subcarrier (B), and PHY and MAC data rates for different MCS types.

In Table 4.6, the calculations of the MAC data rate are obtained for 26 data OFDM symbols using the PUSC permutation mode, where only one type

TABLE 4.6

802.16e DL Subframe Data Rate Analysis

MCS			Slot Capacity (bits)	Dr_{PHY} (Mbps)	Dr_{MAC} (Mbps)
Modulation Type	Code Rate Type (Cr)	B (bits/subcarrier)			
64 QAM	5/6	5	240	43.98	18.72
64 QAM	3/4	4.5	216	31.48	16.84
64 QAM	2/3	4	192	27.98	14.97
64 QAM	1/2	3	144	20.99	11.23
16 QAM	3/4	3	144	20.99	11.23
16 QAM	2/3	2.666667	128	18.65	9.98
16 QAM	1/2	2	96	13.99	7.48
QPSK	3/4	1.5	72	10.49	5.61
QPSK	1/2	1	48	6.99	3.74

of MCS is used for all OFDM symbols. The slot capacity decreases when low modulation order is used and vice versa. In line with this, the data rate score in PHY is higher than that in MAC. This is due to the aggregate data bits in the PHY mode being obtained for one OFDM symbol time (102.9 µs), while the aggregate data bits in the MAC are obtained for a one-frame time period (5 ms).

The data rate calculation in the FFR technique has different aspects. The MAC data rate calculation in Equation 4.12 cannot be used to calculate the data rate in the FFR technique for the following reasons. First, the DL subframe structure is different in FFR than that in WiMAX (without FFR), since the former requires the partitioning of the DL subframe into two zones with different sizes, whereas the latter utilizes the DL subframe as one unit. In addition, FFR serves users by one of the zones based on the location of the user in the cell area, which is not the case in WiMAX without FFR. In FFR, the response of the data rate is different from one zone to another, since zones serve users with different locations, which leads to the use of different modulations and code rate types for each user channel condition. Therefore, the data rate should be calculated per zone, separately. Hence, to calculate the MAC data rate in FFR, the aforementioned reasons need to be addressed. Accordingly, derivation of the MAC data rate for the FFR technique is presented as follows.

The burst profile is a MAC control message that holds the type of MCS. It is used for the link adaptation procedure and is selected based on the user's SINR value. There are several types of MCS defined for different levels of SINR. Mathematically, the MCS can be represented as $B = Cr\log_2(Q)$, where it is the result of Equation 4.13. Given that the number of slots β reserved is different from one user to another, the number of subcarriers Kr_{slot} is the same for each slot. Therefore, a formula to calculate the MAC data rate in the FFR technique can be derived from Equation 4.12 as follows:

$Dr_{MAC}(M)$

$$
= \begin{cases} \dfrac{1}{Tf}\displaystyle\sum_{u=1}^{\alpha}\sum_{i=1}^{\beta}B(u)P(u)Kr_{slot}(i), & \text{for } P_{SINR}^{min}(M) \le SINR(u) \le P_{SINR}^{max}(M) \\[2em] & \text{where } M \in \{R1, R3_A, R3_{BC}\} \\[1em] 0 & \text{elsewhere} \end{cases} \tag{4.14}
$$

where:
- α is the number of active users (actually served) in the target M zone or segment.
- M represents either the R1 zone or one of the segments in the R3 zone.
- Tf is the frame duration time.
- B represents the number of data bits per subcarrier obtained based on the MCS type, and the MCS is selected according to the user channel condition (SINR(u)).

The condition P(u) equals 1 if the active user SINR is within the zone or segment coverage area (P_{SINR}^{min} and P_{SINR}^{max}); otherwise, it equals zero. The zone or segment boundaries are determined by defining suitable SINR thresholds.

In reference to the proposed scenario in Section 4.2.2, the R1 zone is able to serve 60 users with a load of four slots per user, whereas the R3 zone can serve 10 users with load of five slots per user in each segment (A, B, and C). Consequently, after reloading the values of α and β by the commensurate values of the target zone or segment, Equation 4.14 can be extended to calculate the MAC data rate for each of the R1 zone, segment A, and segment BC in the R3 zone, as follows:

$Dr_{MAC}(M)$

$$
= \begin{cases} \dfrac{1}{Tf}\displaystyle\sum_{u=1}^{\alpha}\sum_{i=1}^{\beta}B(u)P(u)Kr_{slot,i}(i), & : M = R1, \quad \alpha = 60, \beta = 4 \\[1.5em] \dfrac{1}{Tf}\displaystyle\sum_{u=1}^{\alpha}\sum_{i=1}^{\beta}B(u)P(u)Kr_{slot,i}(i), & : M = R3_A, \quad \alpha = 10, \beta = 5 \\[1.5em] \dfrac{1}{Tf}\displaystyle\sum_{u=1}^{\alpha}\sum_{i=1}^{\beta}B(u)P(u)Kr_{slot,i}(i), & : M = R3_{BC}, \quad \alpha = 20, \beta = 5 \end{cases} \tag{4.15}
$$

The value of α in Equation 4.15 represents the maximum number of users in the intended zone or segment. Practically, $\alpha \leq 60, 20,$ or 10, where it depends on the number of active users in the target zone or segment. Finally, the MAC data rate of traditional FFR and the proposed FFR models can be calculated as in Equations 4.16 and 4.17, respectively.

$$Dr_{MAC}^{Trd} = Dr_{MAC}(R1) + Dr_{MAC}(R3_A) \text{ bps} \tag{4.16}$$

$$Dr_{MAC}^{Pro} = Dr_{MAC}(R1) + Dr_{MAC}(R3_A) + Dr_{MAC}(R3_{BC}) \text{ bps} \tag{4.17}$$

Equation 4.17 is used to calculate the MAC data rate for both SRA and DRA algorithms. Consequently, the average data rate as a function of the number of trials (Z) can be obtained as follows:

$$\overline{Dr}_{MAC}^{Trd} = \frac{1}{Z} \sum_{i=1}^{Z} Dr_{MAC}^{Trd}(i) \text{ bps} \tag{4.18}$$

$$\overline{Dr}_{MAC}^{Pro} = \frac{1}{Z} \sum_{i=1}^{Z} Dr_{MAC}^{Pro}(i) \text{ bps} \tag{4.19}$$

4.5.3 Subcarrier Efficiency Modeling

Subcarrier efficiency represents the aggregate number of data bits carried by the subcarriers in each burst normalized by the total number of bursts in the intended zone or segment [24]. The normalized subcarrier efficiency (Kr_E) can be obtained as follows:

$$Kr_E(M) = \frac{1}{\alpha\omega} \sum_{u=1}^{\alpha(M)} \sum_{j=1}^{\omega(M)} Cr_{u,j}(M) b_{u,j}(M) \text{ b/subcarrier/burst} \tag{4.20}$$

where:
α is the number of active users in the target zone or segment.
Cr denotes the code rate type.
b is the number of data bits carried by a subcarrier, which corresponds to a particular modulation type.
ω is the number of subcarriers reserved for a specific user load and equals $(\beta \times Kr_{slot})$, where β is the number of slots per user and Kr_{slot} is the number of subcarriers per slot.

Note that the burst has been defined as serving one user with a specific load ($\omega = \beta \times Kr_{slot}$), as mentioned in Section 4.2.2. Therefore, $\alpha\omega$ represents the number of allocated (used) bursts in the target zone or segment, and the units of Equation 4.20 can be attributed per burst. M indicates the frame part type in the DL subframe and may be equal to R1, R3 (segment A), or segment BC. Accordingly, the average subcarrier efficiency is calculated based on the number of trials (Z) as follows:

$$\overline{Kr_E}(M) = \frac{1}{Z}\sum_{i=1}^{Z} Kr_E^M(i)\, b/subcarrier/burst \qquad (4.21)$$

The output of Equation 4.21 represents the average subcarrier efficiency per zone or segment. In order to find the subcarrier efficiency per model, the arithmetic mean is used to find the overall subcarrier efficiency for all frame parts, such as $Kr_E^{Trd} = \left[\overline{Kr_E}(R1) + \overline{Kr_E}(R3_A)\right]/2$ for traditional FFR and $Kr_E^{Pro} = \left[\overline{Kr_E}(R1) + \overline{Kr_E}(R3_A) + \overline{Kr_E}(R3_{BC})\right]/3$ for proposed FFR.

4.5.4 Spectral Efficiency Modeling

The main goal for network designers in cellular network deployment is to increase the coverage area and capacity, and from the capacity standpoint, the most important measure in base station performance is spectral efficiency, which can be considered the production of bandwidth [21]. Spectral efficiency is directly proportional to the system data rate divided by the effective bandwidth (EBW) and the average frequency reuse factor (AFRF), where EBW is the operating bandwidth that is appropriately scaled by the DL/UL ratio [25,26]. If the aggregate data rate (*dr*) of N active users in the DL subframe is divided by EBW and AFRF, then DL spectral efficiency (DL_{SE}) can be defined as follows:

$$DL_{SE} = \frac{\sum_{i=1}^{N_{DL}} dr_{DL}(i)}{EBW \times AFRF}\ bps/Hz \qquad (4.22)$$

In traditional FFR, the FRF in the R3 zone (FRF_{R3}) equals 3 and that in the R1 zone (FRF_{R1}) equals 1 [27]; for instance, the FRF_{R3} in a grid of 19 cells is given in Equation 4.23 [28]. On the other hand, in this work, all the subchannels in the R3 zone are used. As a result, FRF_{R3} equals 1 in the R3 zone, as well as in the R1 zone. Theoretically, this leads to a 66% increase in spectral efficiency in the R3 zone compared with traditional FFR.

$$FRF_{R3} = \frac{1}{\left[\dfrac{\left[6\times\left(\dfrac{1}{3}\right) + 6\times\left(\dfrac{1}{3}\right) + 7\times\left(\dfrac{1}{3}\right)\right]}{19}\right]} = 3 \qquad (4.23)$$

The number of OFDM symbols reserved for R1 or R3 is different from one design to another. Moreover, R1 exploits all the subchannels, whereas R3 exploits only one-third of them. Therefore, the term AFRF is introduced in Equation 4.22 to define a unified FRF based on the reserved OFDM symbols for both the R1 and R3 zones as follows [17]:

$$AFRF = FRF_{R1} \frac{N1}{Nt} + FRF_{R3} \frac{N3}{Nt} \tag{4.24}$$

here $N1$ and $N2$ represent the number of data OFDM symbols in the R1 zone and R3 zone, respectively. In addition, the number of data OFDM symbols in the DL subframe is equal to Nt, where $Nt = N1 + N3$. In conclusion, Equation 4.22 can be used to calculate the spectral efficiency of the traditional FFR algorithm and proposed FFR algorithms when the aggregate data rate of the aforementioned algorithms is involved in the calculations. Finally, the average spectral efficiency can be obtained for the whole test period (Z) in Equation 4.25.

$$\overline{DL_{SE}} = \frac{1}{Z} \sum_{i=1}^{Z} DL_{SE}(i) \text{ bps/Hz} \tag{4.25}$$

4.5.5 FFR DL Subframe Capacity Modeling

DL subframe capacity is defined by two criteria: the number of utilized slots and the number of active users. In the following subsections, these two criteria are discussed and calculated as follows.

4.5.5.1 Slot Utilization Modeling

The WiMAX standard defines the number of subchannels in a given bandwidth, the number of OFDM symbols in a 5 ms WiMAX frame, and the slot dimensions [18]. The slot dimensions depend on the permutation type and can be one subchannel × 1, 2, or more OFDM symbols [18]. Consequently, the number of slots in a WiMAX frame can be computed. Referring to the proposed DL subframe design (see Section 4.2.2), burst has been defined to serve one user with a specific number of slots. The number of utilized slots $N_{slot}(M)$ per zone or segment is calculated in Equation 4.26.

$$N_{slot}(M) = \sum_{i=1}^{\alpha(M)} S_{burst}^{M}(i) : M \in \{R1, R3_A, R3_{BC}\} \tag{4.26}$$

where α represents the number of active users in a particular zone or segment. The parameter M refers to the zone or segment type. The number of slots per burst $S_{\text{burst}}^{\text{M}}$ depends on the burst definition. The burst spans five slots when M equals R3$_A$ or R3$_{BC}$ and four slots when M equals R1. The number of slots in Equation 4.26 must not exceed the maximum capacity of the intended zone or segment. Therefore, the number of slots should satisfy the condition below.

$$N_{\text{slot}}(M) \le N_{\text{subch}}(M) \left\lceil \frac{N_{\text{OFDM}}^{\text{DL}} - N_{\text{OFDM}}^{\text{OH}} - N_{\text{OFDM}}(m)}{\varphi} \right\rceil \tag{4.27}$$

$$: m = \begin{cases} R3_A, & \text{if } M = R1 \\ R1, & \text{if } M = R3_A \\ R1, & \text{if } M = R3_{BC} \end{cases}$$

The right side of Equation 4.27 represents the maximum number of slots per zone or segment. The number of used subchannels N_{subch} is variable depending on the target zone or segment (M). $N_{\text{OFDM}}^{\text{DL}}$ and $N_{\text{OFDM}}^{\text{OH}}$ represent the number of OFDM symbols reserved for the DL subframe and overhead, respectively. $N_{\text{OFDM}}(m)$ denotes the number of OFDM symbols occupied by another part of the frame that is not equal to the current part of the frame (M). φ is the required number of OFDM symbols per slot, which is defined based on the used permutation type and the link direction. The total numbers of used slots by traditional FFR and proposed FFR are given, respectively, as follows:

$$N_{\text{slot}}^{\text{Trd}} = \sum_{M \in \{R1, R3_A\}} N_{\text{slot}}(M) \tag{4.28}$$

$$N_{\text{Slot}}^{\text{Pro}} = \sum_{M \in \{R1, R3_A, R3_{BC}\}} N_{\text{slot}}(M) \tag{4.29}$$

In order to compare the performance of the traditional and proposed FFR models for the entire test period (Z), the average numbers of slots for the aforementioned models are calculated as follows:

$$\overline{N}_{\text{slot}}^{\text{Trd}} = \frac{1}{Z} \sum_{i=1}^{Z} N_{\text{slot}}^{\text{Trd}}(i) \tag{4.30}$$

$$\overline{N}_{\text{slot}}^{\text{Pro}} = \frac{1}{Z} \sum_{i=1}^{Z} N_{\text{slot}}^{\text{Pro}}(i) \tag{4.31}$$

4.5.5.2 Active User Modeling

Users are randomly distributed in the target cell, and the number of users who are in any part of the cell area is a random variable. Referring to the proposed design, the area of operation for each zone or segment is defined by appropriate SINR thresholds. Therefore, the number of active users per zone or segment can be calculated as below.

$$
N_{\text{user}}(M)
$$

$$
= \begin{cases} \displaystyle\sum_{u=1}^{\infty(M)} \text{User}(u), & : \text{if } P_{\text{SINR}}^{\min}(M) \leq \text{SINR}(u) \leq P_{\text{SINR}}^{\max}(M) \\[2em] & \text{where, } M \in \{R1, R3_A, R3_{BC}\} \\[1em] 0, & \text{elsewhere} \end{cases} \tag{4.32}
$$

where User(u) is a user indexed. Equation 4.32 calculates the number of active users (∞) in a certain zone or segment (M) as long as the SINR of the intended user is within the zone or segment boundaries. The boundaries of the zone or segment are represented by upper and lower SINR thresholds, which are denoted as ($P_{\text{SINR}}^{\min}$ and $P_{\text{SINR}}^{\max}$). The synonym interpretation to the number of active users is the number of utilized slots. From the standpoint of design, the number of users reflects the number of used slots. However, it is possible that the number of active users exceeds the maximum capacity of user numbers per zone or segment. Therefore, Equation 4.32 must satisfy the condition in Equation 4.33.

$$
N_{\text{user}}(M) \leq \frac{N_{\text{slot}}^{\max(M)}}{\beta(M)} : \begin{cases} \text{if } M = R1, & N_{\text{slot}}^{\max} = 240, \beta = 4 \\ \text{if } M = R3_A, & N_{\text{slot}}^{\max} = 50, \beta = 5 \\ \text{if } M = R3_{BC}, & N_{\text{slot}}^{\max} = 100, \beta = 5 \end{cases} \tag{4.33}
$$

$N_{\text{slot}}^{\max}$ represents the maximum number of slots specified for a certain zone or segment (M). The maximum number of slots can be deduced from the definition of WiMAX system parameters reported in IEEE standards [18]. β refers to user load (slots) and is specified according to the type of zone or segment. Consequently, the right side of Equation 4.33 represents the maximum number of users that can be served by a zone or segment. The total number of active users in traditional FFR and proposed FFR models can be obtained in Equations 4.34 and 4.35, respectively, as follows:

$$
N_{\text{user}}^{\text{Trd}} = \sum_{M \in \{R1, R3_A\}} N_{\text{user}}(M) \tag{4.34}
$$

$$
N_{\text{user}}^{\text{Pro}} = \sum_{E \in \{R1, R3_A, R3_{BC}\}} N_{\text{user}}(M) \tag{4.35}
$$

Finally, the average number of users in the traditional and proposed FFR models is obtained for the whole test period (Z) and calculated as in Equations 4.36 and 4.37, respectively.

$$\bar{N}_{user}^{Trd} = \frac{1}{Z} \sum_{i=1}^{Z} N_{user}^{Trd}(i) \qquad (4.36)$$

$$\bar{N}_{user}^{Pro} = \frac{1}{Z} \sum_{i=1}^{Z} N_{user}^{Pro}(i) \qquad (4.37)$$

4.5.6 Channel Capacity Modeling

Channel capacity is an important metric used to evaluate the performance of wireless systems. The capacity determines the amount of information (data) that can be reliably delivered to and from users in the communication system, where every wireless channel has a limit as to how much data can be reliably sent [29]. In 1948, Claude Shannon found a formula to calculate the limits of reliable communication channels. The Shannon capacity formula in cellular systems, applying the FRF, is shown in Equation 4.38 [30]:

$$C = BW\, FRF\, \log_2(1 + SINR)\, bits/s \qquad (4.38)$$

FRF refers to the frequency planning technique (see Section 2.4.1), where the available bandwidth is divided by the number of cells in the grid. This is different than our case, since using the FFR technique obviates the need for frequency planning [2]. The FRF in FFR has two values; in the R1 zone it is equal to 1, and in the R3 zone it is equal to 3. Moreover, the bandwidth utilization in the WiMAX base station under FFR has different aspects. The OFDMA technique helps to divide the bandwidth into a number of subchannels, and the number of these subchannels depends on the used system bandwidth, the type of subcarrier permutation, and the transmission direction (DL or UL) [31]. Besides, in FFR the number of used subchannels is different in the R1 zone than in the R3 zone (segment A), as well as in segment BC [2]. Therefore, Equation 4.38 needs to be modified to meet FFR technique requirements, as below.

$$C_{FFR} = EBW\, \delta\, \rho\, \log_2(1 + SINR)\, bits/s \qquad (4.39)$$

where EBW is the effective bandwidth, the same as in Equation 4.22. δ denotes the ratio of utilized bandwidth or subchannels ($\delta = N_{subch}(M)/N_{subch}$), where $N_{subch}(M)$ represents the number of assigned subchannels to a particular zone or segment (M) and N_{subch} is the total number of available subchannels in the system. The value of N_{subch} depends on the system parameters. ρ is the ratio of data OFDM symbols used by a particular zone or segment to the total number of data OFDM symbols occupied in the DL subframe. If the channel capacity is calculated for a system using universal frequency (FRF = 1), then δ and ρ are equal to 1. However, this book focuses on the DL subframe, and the DL subframe is divided into two parts, the R1 and R3 zones. Therefore, Equation 4.39 needs to be divided into two parts as in [32], since the bandwidth utilization in the R1 and R3 zones is unequal.

In order to find the capacity of traditional FFR and the proposed SRA and DRA algorithms, assuming a user with load is equal to the visible subchannels in the target zone or segment, and this user moves from the base station toward the cell border, Equation 4.39 can be rewritten as follows:

$$
C_{FFR}^{Trd} = \begin{cases} EBW\ \delta\rho\ \log_2(1+SINR), & : \quad \delta = 1, \\[1.5ex] \rho = \dfrac{N1}{Nt}, \text{if R1 zone} \leftarrow \text{resource assignment} \\[1.5ex] BW\ \delta\rho\ \log_2(1+SINR), & : \quad \delta = \dfrac{1}{3}, \\[1.5ex] \rho = \dfrac{N3}{Nt}, \text{if R3 zone} \leftarrow \text{resource assignment} \\[1.5ex] 0, & \quad \text{elsewhere} \end{cases} \tag{4.40}
$$

$$
C_{FFR}^{Pro} = \begin{cases} EBW\ \delta\rho\ \log_2(1+SINR), & : \quad \delta = 1, \\[1.5ex] \rho = \dfrac{N1}{Nt}, \text{if R1 zone} \leftarrow \text{resource assignment} \\[1.5ex] EBW\ \delta\rho\ \log_2(1+SINR), & : \quad \delta = 1, \\[1.5ex] \rho = \dfrac{N3}{Nt}, \text{if R3 zone} \leftarrow \text{resource assignment} \\[1.5ex] 0, & \quad \text{elsewhere} \end{cases} \tag{4.41}
$$

where N_1, N_3, and N_t are the same as in Equation 4.24. The capacity calculation in Equations 4.40 and 4.41 illustrates the capacity per zone. Therefore, the benefit of using the proposed algorithms can be observed by comparing the results of these equations with each other. Moreover, when $\delta = 2/3$, Equation 4.41 illustrates the gain of using segment BC in the R3 zone.

4.6 Performance Evaluation of FFR Models versus Traditional FFR

In this section, different methods are used to evaluate the work done in this book. The purpose of these methods is to view the performance evaluation of the proposed algorithms in comparison with traditional FFR in different formats, as follows.

4.6.1 Evaluation by Metrics

All the metrics given in Section 4.5 are used to evaluate and compare the performance of the proposed SRA and DRA algorithms with the traditional FFR as follows:

1. *Data rate*: Data rate is the amount of data transferred in a particular channel condition during a specific time period. This metric is used to show that the new algorithms have the ability to increase the data rate of the traditional FFR under the same channel condition and time period.

2. *Subcarrier efficiency*: Subcarrier efficiency shows how many data bits each frame part holds. It assists in exploring the response of the DL subframe parts of the proposed models and compares them with the DL subframe parts of the existing technology (traditional FFR).

3. *Spectral efficiency*: Spectral efficiency is the amount of information transmitted over a given bandwidth in a certain communication system. It indicates how efficiently a limited bandwidth is utilized by the MAC layer. This metric is used to compare the performance of the traditional FFR and the proposed SRA and DRS algorithms to show how efficiently the MAC layer operates under the new design.

4. *DL subframe capacity*: Frame capacity refers to two metrics, the number of utilized slots and the number of active users. These metrics are presented to guarantee that the proposed algorithms increase the number of active users and resource utilization in a given frame duration. Specifically, these metrics show how efficiently the DL subframe is exploited under the new algorithms.

5. *Channel capacity*: The channel capacity indicates how efficiently the available bandwidth is exploited, since the results of this metric are directly commensurate with the number of used subchannels and FRF proportion. This metric is used to ensure that all the available subchannels are utilized in the proposed SRA and DRA algorithms,

which in return leads to an enhancement in system performance. This is unlike the traditional FFR, where only 33% of the subchannels are used in the R3 zone when PUSC is applied.

4.6.2 Evaluation by Different Configurations

In order to test the performance of the proposed SRA and DRA algorithms, different configurations are used to show the robustness of these algorithms in different contexts, as follows:

1. *Static and dynamic FFR*: As stated in Section 2.5, FFR can be implemented using static or dynamic configurations. In this book, these two types of configuration are considered when designing the SRA FFR and DRA FFR algorithms, respectively. The dynamic configuration is used to overcome the problem of variation in population density. Moreover, two mobility patterns were proposed to show the benefits of the dynamic algorithm (DRA), which are the random mobility pattern (DRA-I), where the users are randomly dropped in the cell coverage area, and the directed mobility pattern (DRA-II), where groups of users are moved from the base station toward the cell center edge.

2. *System stability*: The proposed SRA and DRA algorithms are implemented using different system parameters. Mobile WiMAX can operate in different sets of system parameters, as mentioned in Section 3.2. In order to highlight the stability of the proposed algorithms, three types of WiMAX system parameters are considered: 5, 10, and 20 MHz system bandwidth.

4.7 Performance Evaluation of Proposed FFR Models versus Other Related Works

As IEEE 802.16e is a multiple access scheme build on the OFDMA technique, different scenarios of resource allocation can be implemented. Besides, IEEE 802.16e can be run in different types of system parameters, which results in different performance. For instance, changing the number of OFDM symbols for the DL subframe, R1 zone, or R3 zone leads to different performance. These facts should be considered when comparing the results of the proposed algorithms with others. However, in order to validate the results of SRA and DRA algorithms, a comparison with other related work is performed in this book.

4.8 Summary

In cellular networks and when using universal frequency (FRF of 1), the large demand of E-services will be satisfied, but users of cellular networks will suffer from ICI, especially at the cell border. Several techniques have been proposed to address the ICI problem; one of these is the FFR technique. Most of these techniques are built on the basis of using part of the bandwidth in each cell to mitigate the ICI effect. However, one of the drawbacks of the FFR technique is the inefficient utilization of bandwidth and resources in the DL subframe. This problem has been tackled in this book where a universal frequency is achieved while maintaining the ICI effect at an acceptable level.

In this chapter, the design, modeling, network scenarios, and system parameters, as well as the metrics that are used to assess the performance of the novel algorithms SRA FFR and DRA FFR, are presented. The methodology in this chapter starts by analyzing the DL subframe parts in an interference environment to better exploit these parts, and then designing the network environment and setting the necessary base station parameters. The SRA FFR is proposed to enhance the performance of traditional FFR in terms of number of served users, resource utilization, data rate, and spectral efficiency. Although the SRA algorithm enhances the performance of traditional FFR, it cannot address the constantly changing population density, since the SAR is statically configured and the configuration remains unchanged for a period of time. Therefore, DRA FFR is proposed to overcome the variations in population density and at the same time enhance the performance of traditional FFR through the same metrics that are considered in SRA FFR. In both proposed algorithms, all the available subchannels are used, unlike traditional FFR, where only one-third of subchannels are used in the R3 zone. Therefore, the wastage of resources and bandwidth is handled by the proposed new algorithms.

Selective metrics and methods are used to show the efficiency of these algorithms in different conditions, such as using both static and dynamic configurations, using different system bandwidths, and using different mobility patterns. In Chapter 5, a discussion of the results related to the proposed SRA and DRA algorithms is presented.

References

1. A. Goldsmith, *Wireless Communications*, Cambridge University Press, Cambridge, 2005.
2. Forum, Mobile WiMAX—Part I: A technical overview and performance evaluation, 2006, p. 53.

3. Forum, WiMAX™ system evaluation methodology, version 2.1, 2008, p. 209.
4. P. Scully, J. Nelson, S. Mcgrath, J. Johnson, R. Skehill, and E. Cano, Mobility in wireless communication networks, in World Scientific Publishing Company, Sudip Misra (Author, Ed.), Subhas C. Misra (Ed.), Isaac Woungang (Ed.) *Selected Topics in Communication Networks and Distributed Systems*, Singapore, 2010, pp. 1–41.
5. K. R. Manoj, *Coverage Estimation for Mobile Cellular Networks from Signal Strength Measurements*, University of Texas, Dallas, 1999.
6. M. C. Necker, A novel algorithm for distributed dynamic interference coordination in cellular OFDMA networks, doctor-engineer, Faculty of Computer Science, Electrical Engineering and Information Technology, Stuttgart, 2009. http://www.ikr.uni-stuttgart.de/Content/Publications/Archive/Ne_Diss_36817.
7. Forum, Mobile WiMAX—Part II: A comparative analysis, 2006, p. 47.
8. R. Giuliano, C. Monti, and P. Loreti, WiMAX fractional frequency reuse for rural environments, *IEEE Wireless Communications*, vol. 15, pp. 60–65, 2008.
9. H. Lei, X. Zhang, and D. Yang, A novel frequency reuse scheme for multi-cell OFDMA systems, in *2007 IEEE 66th Vehicular Technology Conference, VTC Fall 2007*, Canada, 2007, pp. 347–351.
10. M. Maqbool, P. Godlewski, M. Coupechoux, and J.-M. Kélif, Analytical performance evaluation of various frequency reuse and scheduling schemes in cellular OFDMA networks, *Performance Evaluation*, vol. 67, pp. 318–337, 2010. www.sciencedirect.com/science/article/pii/S0166531609001011?via%3Dihub.
11. S. R. Boddu, A. Mukhopadhyay, B. V. Philip, and S. S. Das, Bandwidth partitioning and SINR threshold design analysis of fractional frequency reuse, in *National Conference on Communications (NCC)*, 2013, pp. 1–5.
12. A. Ganz, Z. Ganz, and K. Wongthavarawat, *Multimedia Wireless Networks: Technologies, Standards and QoS*, Prentice Hall, Upper Saddle River, NJ, 2004.
13. S.-I. Chakchai, R. Jain, and A. K. Tamimi, Scheduling in IEEE 802.16e mobile WiMAX networks: Key issues and a survey, *IEEE Journal on Selected Areas in Communications*, vol. 27, pp. 156–171, 2009.
14. I. WP8F, Additional technical details supporting IP-OFDMA as an IMT-2000 terrestrial radio interface, Contribution ITU-R WP/1079, 2006.
15. L. Nuaymi, *WiMAX: Technology for Broadband Wireless Access*, Wiley, Hoboken, NJ, 2007.
16. C. So-In, R. Jain, and A.-K. Tamimi, Capacity evaluation for IEEE 802.16 e mobile WiMAX, *Journal of Computer Systems, Networks, and Communications*, vol. 2010, p. 1, 2010.
17. I. N. Stiakogiannakis, G. E. Athanasiadou, G. V. Tsoulos, and D. I. Kaklamani, Performance analysis of fractional frequency reuse for multi-cell WiMAX networks based on site-specific propagation modeling [wireless corner], *IEEE Antennas and Propagation Magazine*, vol. 54, pp. 214–226, 2012.
18. IEEE-Std, IEEE 802.16–2009: IEEE standard for local and metropolitan area networks part 16: Air interface for broadband wireless access systems, IEEE Standards (revision of IEEE Standard 802.16–2004), 2009, pp. 1–2080.
19. F. P. Font)n and P. M. Espiñeira, *Modelling the Wireless Propagation Channel: A Simulation Approach with MATLAB*, vol. 5, Wiley, Hoboken, NJ, 2008.
20. S. Glisic and B. Lorenzo, *Advanced Wireless Networks: Cognitive, Cooperative and Opportunistic 4G Technology*, Wiley, Hoboken, NJ, 2009.

21. J. G. Andrews, A. Ghosh, and R. Muhamed, *Fundamentals of WiMAX: Understanding Broadband Wireless Networking*, Pearson Education, London, 2007.
22. A. Kumar, *Mobile Broadcasting with WiMAX: Principles, Technology, and Applications*, Taylor & Francis, Boca Raton, FL, 2008.
23. D. Pareit, V. Petrov, B. Lannoo, E. Tanghe, W. Joseph, I. Moerman, et al., A throughput analysis at the MAC layer of mobile WiMAX, in *2010 IEEE Wireless Communications and Networking Conference (WCNC)*, Sydney, Australia, 2010, pp. 1–6.
24. J. J. J. Roy and V. Vaidehi, Analysis of frequency reuse and throughput enhancement in WiMAX systems, *Wireless Personal Communications*, vol. 61, pp. 1–17, 2011.
25. S. Ahmadi, *Mobile WiMAX: A Systems Approach to Understanding IEEE 802.16 m Radio Access Technology*, Academic Press, Amsterdam, 2011.
26. Y. Gao, X. Zhang, Y. Jiang, and J.-W. Cho, System spectral efficiency and stability of 3G networks: A comparative study, in *IEEE International Conference on Communications, ICC'09*, Dresden, Germany, 2009, pp. 1–6.
27. A. Darwish, A. S. Ibrahim, A. H. Badawi, and H. Elgebaly, Performance improvement of fractional frequency reuse in WiMAX network, in *2011 IEEE 73rd Vehicular Technology Conference, VTC Spring 2011*, Budapest, Hungary, 2011, pp. 1–5.
28. R. Ghaffar and R. Knopp, Fractional frequency reuse and interference suppression for OFDMA networks, in *Proceedings of the 8th International Symposium on Modeling and Optimization in Mobile, Ad Hoc and Wireless Networks (WiOpt)*, Avignon, France 2010, pp. 273–277.
29. K. Dietze, T. Hicks, and G. Leon, WiMAX system performance studies, EDX Wireless, Eugene, OR, 2011.
30. D. Tse and P. Viswanath, *Fundamentals of Wireless Communication*, Cambridge University Press, Cambridge, 2005.
31. L. Korowajczuk, *LTE, WIMAX and WLAN Network Design, Optimization and Performance Analysis*, Wiley, Hoboken, NJ, 2011.
32. P. Godlewski, M. Maqbool, M. Coupechoux, and J.-M. Kélif, Analytical evaluation of various frequency reuse schemes in cellular OFDMA networks, in *Proceedings of the 3rd International Conference on Performance Evaluation Methodologies and Tools*, Greece, 2008, p. 32.

5

Performance Analysis and Evaluation of Proposed FFR Algorithms

---

5.1 Introduction

Wireless communication has grown quickly in the last decades. It has captured the attention of the media and the communications industry, especially mobile communications technology. Most modern devices depend on wireless communication to perform their jobs. These devices vary according to the nature of the device specifications and the purpose for which they are used. In point-to-multipoint (PMP) topology, mobile devices must be connected to a focal point called a base station [1]. There are many types of mobile base stations on the market. To choose from any of them, one needs to study different types of criteria to evaluate their performance. Some criteria depend on numerical results, measuring, comparing quantities, analyzing measurements, and so on. What compounds the problem even more, and especially in the case of WiMAX, is that the IEEE 802.16e base stations can operate on different configuration settings. Each configuration has different sets of system parameters. For this reason, it is important to define certain criteria to evaluate the performance of these base stations, such as bandwidth utilization, number of served users, resource utilization, data rate, and spectral efficiency (SE).

The performance evaluation of the static resource assignment (SRA) and dynamic resource assignment (DRA) algorithms are presented in this chapter in order to justify the decision to use these algorithms in mobile WiMAX base stations.

The rest of this chapter is organized as follows: The performance analysis of the SRA fractional frequency reuse (FFR) algorithm is introduced in Section 5.2. The response of segment BC through the proposed cases is presented in Section 5.2.1, and the results of SRA FFR and traditional FFR are compared in Section 5.2.2. In Section 5.3, the performance analysis of the DRA FFR algorithm is presented, where two types of mobility patterns are tested: mobility pattern I (DRA-I) and mobility pattern II (DRA-II).

In Section 5.3.1, the performance analysis of DRA-I FFR is introduced. First, the response of segment BC through the proposed layers is presented; then it is followed by comparing the results of DRA-I FFR with those of traditional FFR. However, the performance analysis of second mobility (DRA-II FFR) is introduced in Section 5.3.2. It starts by analyzing the response of segment BC through the proposed layers, and is then followed by a comparison of the results of DRA-II with those of traditional FFR. All the results of SRA and DRA algorithms are compared with traditional FFR in Section 5.4. Then in Section 5.5, selective metrics are used to evaluate the robustness of SRA and DRA algorithms under different system bandwidths. In addition, this section includes the results of channel capacity per SRA and DRA in different system bandwidths. In order to validate the results of the proposed algorithms, Section 5.6 shows the results' comparisons of SRA and DRA algorithms with other related works. Finally, the work done in this chapter is summarized in Section 5.7.

5.2 Performance Analysis and Evaluation of the SRA FFR Algorithm

The response of the SRA algorithm is evaluated under different coverage areas of segment BC. As in [2,3], four cases are proposed (see Section 4.3.1) to find the best performance (coverage area) for segment BC. The evaluation process is divided into two parts. The first part represents the performance response in each case, and the second part compares the response of SRA FFR with that of traditional FFR. However, our interest in the first part is to make a decision about which one of the four cases is suitable as a candidate for WiMAX base station deployment. In contrast, our interest in the second part is to show the benefits of using the SRA FFR algorithm compared with traditional FFR. All metrics used here are on the basis of the analysis presented in Section 4.5. These metrics are used to show the response of the SRA algorithm as follows.

5.2.1 Performance Analysis per Case

The four cases perform differently according to the specifications of each one. The case specifications can be defined as the number of active users in each case, the type of modulation and coding rate used in each case, and the achieved data rate. These specifications are affected by the coverage area in each case. The performances of these cases are explored as follows.

5.2.1.1 Based on Active Users

The users are randomly distributed in the cell coverage area. The number of users located in each case depends on the case coverage area. The criterion used to host users in any case is the threshold signal-to-interference-plus-noise ratio (SINR) mentioned in Table 4.2 (see Section 4.3.1), and these thresholds lead to different areas of coverage. Therefore, the number of active users in each case is different, as shown in Figure 5.1.

The process of hosting users in any case is achieved by applying phase 2 of the SRA algorithm, as mentioned in Section 4.3.2, and the number of active users per case for each trial in Figure 5.1 is the result of Equation 4.32. Taking into account that segment BC can serve a maximum number of 20 users, as mentioned in Section 4.2.2, the results illustrate that cases 3 and 4 can serve more users than other cases. In accordance with the random distribution of users in the system, it is clear that the population density of users in cases 3 and 4 is higher than that in the other cases. Nevertheless, all these cases lose some users since they lie outside their coverage areas. For example, case 2 serves the lowest number of users compared with all cases, whereas case 1 serves more users than case 2.

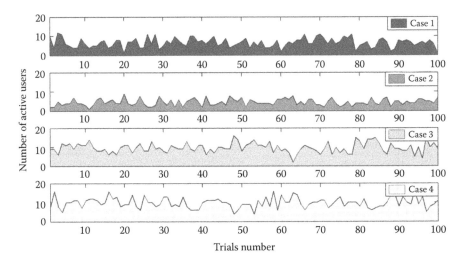

FIGURE 5.1
Number of active users per case.

5.2.1.2 Based on Subcarrier Efficiency

Subcarrier efficiency shows how efficiently the zone or segment (or a case) operates under the new design. It is the missing link for analyzing the response of the downlink (DL) subframe parts (zone or segment). For example, knowing the number of active users and their data rate means that it can specify which zone or segment has the best response. However, how does this zone or segment get this response? The answer to this question is given by calculating the subcarrier efficiency for the intended zone or segment. Figure 5.2 depicts the subcarrier efficiency of segment BC in the four cases. The response of each case is based on Equation 4.20. The SINR thresholds in case 1 are the highest thresholds between cases (Table 4.2), which means that this case uses the highest modulation order, such as 64 QAM with 5/6 coding rate or 64 QAM with 3/4 coding rate, as a result of small path loss. Therefore, the subcarriers in case 1 can carry more data bits than the other cases (maximum 5 bits and minimum 4.5 bits per subcarrier). This interpretation can be repeated for the remaining cases. For instance, the SINR thresholds in case 3 allow the use of a modulation order of 16 QAM with 3/4 coding rate or 16 QAM with 1/2 coding rate. Thus, it holds fewer data bits within its subcarriers than case 1 (maximum 3 bits and minimum 2 bits per subcarrier), as revealed in Figure 5.2. Consequently, it makes sense that the case that carries more data bits can produce the highest data rate, as shown in the following section.

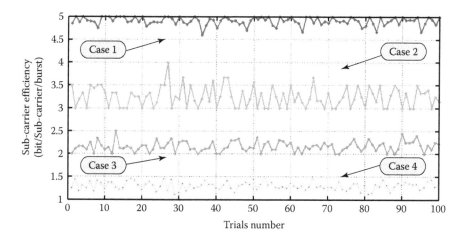

FIGURE 5.2
Subcarrier efficiency per case.

5.2.1.3 Based on Data Rate

Data rate is an important metric used to calculate the data transmission in a given bandwidth. The data rates of the four cases are computed using Equation 4.15 (when M equals to $R3_{BC}$), as shown in Figure 5.3.

In spite of case 1 serving fewer users compared with cases 3 and 4, it shows the highest level of data rate since it holds more data bits within its subcarriers than the other cases (Figure 5.2). Despite case 4 being able to serve more users than cases 2 and 1, and an almost equal number of users in case 3, it shows the lowest data rate among the four cases. This is because case 4

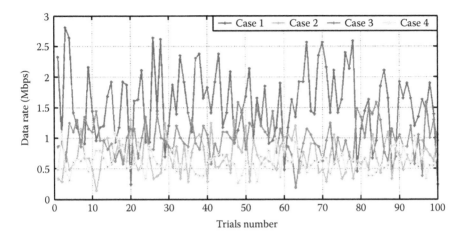

FIGURE 5.3
Data rate response per case.

carries the lowest number of data bits within its subcarriers. However, the responses of cases 2 and 3 are controversial. Case 2 holds more data bits than case 3 but has a lower data rate than case 3. The reason is that case 2 serves far fewer users than case 3, so the number of active users can be considered a compensating factor leading to an increase in the data rate in a case.

The average data rate per case is shown in Figure 5.4. The response of cases 1 and 3 achieves the highest data rate since case 1 uses the highest modulation order type, while case 3 can serve more users than the other cases. In contrast, case 4 can serve more users than case 2 but has a lower data rate than case 2. This is because case 4 uses the lowest modulation order among other cases.

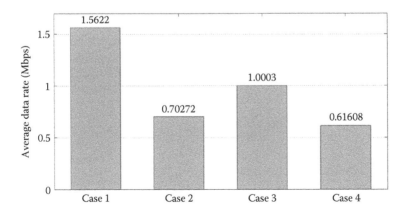

FIGURE 5.4
Average data rate response per case.

The performance of segment BC in the four cases that are set out in Figures 5.1 through 5.4 leads to the following conclusions:

1. The locations of users in each case produce a different distance from the base station. As a result, users experience different channel quality (different SINRs), which leads to a different response per case. This difference in response is a result of using different types of modulation and coding rates.

2. According to number 1, the coverage area of a case near the base station, such as case 1, enables the base station to use the highest modulation order, which will enhance its performance. On the other hand, the coverage area of a case far away from the base station, such as case 4, reduces the possibility of using a high modulation order, which results in poor base station performance.

3. The previous interpretations of numbers 1 and 2 should be linked with the availability of users in the coverage area of the underlining case. The random distribution of users makes their presence different from case to case; this in turn leads to a different response for each case. In other words, case 1 serves more users than case 2, whereas case 2 serves the lowest number of users among all cases. Cases 3 and 4 have the ability to serve more users than cases 1 and 2.

Table 5.1 summarizes the results of these four cases in terms of number of active users, subcarrier efficiency, and data rate, where four levels of performance are introduced to evaluate the response of the four cases, such as highest (HS), high (H), low (L), and lowest (LS). For instance, the case that can achieve the highest data rate compared with the other cases will be referred to as HS.

Analyzing Table 5.1 leads to a trade-off study between cases. Case 1 achieves the highest data rate since it can carry more data bits in its subcarriers. In contrast, case 3 can serve the highest number of users compared with the other cases, in addition to achieving the second highest level of data rate. However, cases 2 and 4 did not achieve the highest (HS) level of performance in terms of the three metrics listed in Table 5.1. Accordingly, the results obtained from cases 1 and 3 are more interesting than the others. Therefore,

TABLE 5.1

Performance Evaluation Summary per Case

Case Number	No. of Active Users	Data Rate	Subcarrier Efficiency
Case 1	L	**HS**	**HS**
Case 2	LS	L	H
Case 3	**HS**	H	L
Case 4	H	LS	LS

in the rest of this book the SRA FFR model will be presented through two algorithms (models): SRA-Case1, where the results of case 1 are considered in the calculation, and SRA-Case3, where the results of case 3 are considered in the calculation.

5.2.2 Performance Comparison between Traditional FFR and SRA FFR

In this section, the performance of traditional FFR is compared with that of the SRA-Case1 and SRA-Case3 algorithms in terms of a variety of metrics, as follows.

5.2.2.1 Comparison Based on Active Users

The average number of active users of traditional FFR (as a result of Equation 4.36), SRA-Case1, and FRA-Case3 (as a result of Equation 4.37) is depicted in Figure 5.5.

As explained in the proposed design (see Section 4.2.2), the traditional FFR model can serve a maximum of 70 users as an average value; of these, 10 are served by the R3 zone and 60 are served by the R1 zone. Traditional FFR works properly when it is capable of serving the maximum number of users (70). This is because the R1 zone and segment BC share the same coverage area (cell center area), and according to the SRA phase 1 algorithm, segment BC starts to serve users after the R1 zone is completely filled by users. Therefore, traditional FFR serves the maximum number of users.

In terms of design, segment BC can serve a maximum of 20 users. Segment BC in the SRA-Case1 model serves only six users per frame, whereas it serves nine users in the SRA-Case3 model. The difference in the number of active users in cases 1 and 3 is related to the availability of users in each case

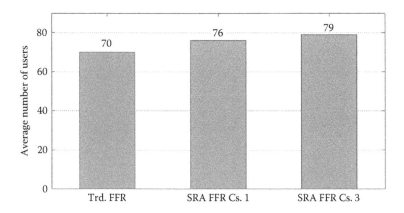

FIGURE 5.5
Average number of active users in traditional FFR and SRA FFR models.

(Figure 5.1). The increment in the number of active users is the result of using phase 2 of the SRA algorithm. However, segment BC misses some users as they are located outside its coverage area. The number of users located outside the coverage area of case 1 is more than that of case 3, since the availability of users in the former area is lower than that in the latter.

5.2.2.2 Comparison Based on Resource Utilization

Resources in wireless communication are always precious. Resource utilization is an important metric used to show the resource exploitation enhancement in the SRA algorithm. Figure 5.6 illustrates the average resource utilization in traditional FFR (including the R1 and R3 zones), SRA-Case1, and SRA-Case3 as a result of Equations 4.28 and 4.29, respectively. As mentioned in Section 4.2.2, traditional FFR can use 290 slots, whereas 50 slots are exploited by the R3 zone and 240 slots are exploited by the R1 zone. Traditional FFR can use all the 290 slots, since it can serve users as much as its capacity allows, as shown in Figure 5.5. On the other hand, the maximum number of slots that can be used in segment BC is 100. The number of utilized slots is increased to 320 and 335 when SRA-Case1 and SRA-Case3 are applied, respectively.

This enhancement of resource utilization is the result of procedures taken to manage the allocation of resources in phase 2 of the SRA algorithm (see Section 4.3.2). Steps 1–10 show the resource assignment in case 1, and steps 21–30 show the resource assignment in case 3. However, as case 3 can serve more users than case 1, case 3 exploits more slots to serve its users. Despite the fact that there are unused slots in segment BC in cases 1 and 3, the SRA algorithm increases resource utilization in traditional FFR by about 10.34% $(((320-290)/290)\times100\%)$ and 15.51% $(((335-290)/290)\times100\%)$ when SRA-Case1 and SRA-Case3 are applied, respectively.

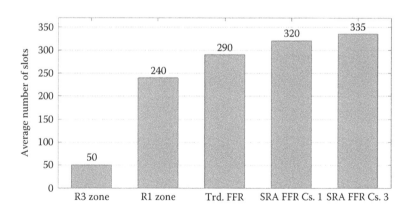

FIGURE 5.6
Average number of slot utilizations in traditional FFR and SRA FFR models.

5.2.2.3 Comparison Based on Subcarrier Efficiency

The subcarrier efficiency of each part of the DL subframe is shown in Figure 5.7, based on Equation 4.20. The response interpretation of cases 1 and 3 is as mentioned in Section 5.2.1.2. The output of the R1 zone represents the aggregate data bits in the subcarriers of used bursts normalized by the total number of bursts, which corresponds to the production of 60 users in the R1 zone. The 60 users represent the maximum number of users that can be served in the R1 zone. Besides, the coverage area of the R1 zone equals the cell center area and uses eight types of burst profiles (Table 4.3). Thus, it carries more data bits than case 3 and the R3 zone, since case 3 and the R3 zone serve fewer users and use a lower modulation order type than that of the R1 zone.

In Figure 5.7, case 3 and the R3 zone almost serve the same number of users (10 in R3 and 9 in case 3) and use convergent modulation and coding scheme (MCS) types (cell border users take advantage of applying FFR). Therefore, case 3 and the R3 zone show a nearly identical performance, as revealed in Figure 5.8, where the average subcarrier efficiency for each part of the DL subframe is calculated as a result of using Equation 4.21. In Figure 5.8, case 1 achieves the highest average value among other results, since it serves users near the base station. Moreover, these users use the highest modulation and coding rate types (Table 4.2) in the system as a result of low path loss, which enables this case to carry more data bits in its subcarriers.

From the standpoint of the available slots, it is helpful to compare the achieved subcarrier efficiency with the number of reserved slots in the R1 zone and segment BC, since the R1 zone uses the largest amount of resources (240 slots) in the DL subframe. In the DL subframe, the number of slots reserved for segment BC (100) represents 41.6% (100/240 = 0.416) of

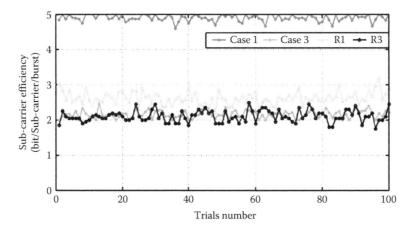

FIGURE 5.7
Subcarrier efficiency response of DL subframe parts in SRA FFR.

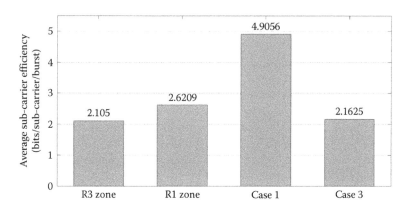

FIGURE 5.8
Average subcarrier efficiency per DL subframe part.

the number of slots reserved for the R1 zone. The subcarrier efficiency of segment BC in case 1 represents 187.21% (4.905/2.620 = 1.8721), and in case 3, 82.51% (2.162/2.620 = 0.8251) of the subcarrier efficiency in the R1 zone. These values of subcarrier efficiency show the effectiveness of using segment BC in the proposed design.

5.2.2.4 Comparison Based on Data Rate

The data rate response of the traditional FFR and the two models (SRA-Case1 and SRA-Case3) can be obtained from Equations 4.16 and 4.17, respectively, as illustrated in Figure 5.9. The SRA-Case1 and SRA-Case3 models show a higher data rate than that of traditional FFR, as a result of using segments B and C in the DL subframe. However, SRA-Case1 has a higher data rate

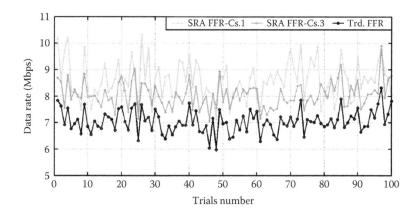

FIGURE 5.9
Data rate response per traditional FFR and SRA FFR models.

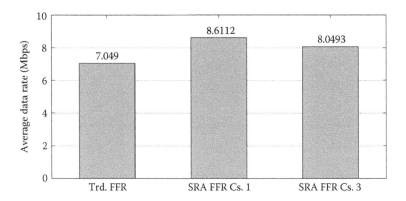

FIGURE 5.10
Average data rate per traditional FFR and SRA FFR models.

than SRA-Case3. This is because SRA-Case1 serves users near the base station through case 1, which leads to the use of high modulation order as a result of low path loss, whereas SRA-Case3 serves users far away from the base station through case 3, which limits the ability of SRA-Case3 to use low modulation order as a result of large path loss. This is because the large path loss reduces the SINR value, which leads to the use of low modulation order.

The average data rates of traditional FFR, SRA-Case1, and SRA-Case3 are obtained by using Equations 4.18 and 4.19, respectively, as shown in Figure 5.10.

The results in Figure 5.10 confirm the results of Figure 5.9 and reveal the importance of exploiting segment BC when using the traditional FFR technique in WiMAX network deployment. As an average value, the SRA-Case3 increases the data rate to about 1 Mbps, while SRA-Case1 increases the data rate to more than 1.5 Mbps.

5.2.2.5 Comparison Based on Spectral Efficiency

Spectral efficiency indicates how efficiently the available bandwidth is utilized. Figure 5.11 illustrates the delineation of the spectral efficiency of the traditional FFR, SRA-Case1, and SRA-Case3 models as a result of using Equation 4.22.

The outputs of the three models in Figure 5.11 are directly proportional to their data rates and the utilized bandwidth. It is worth mentioning that SRA-Case1 and SRA-Case3 attain higher data rates than traditional FFR, as well as using all the available bandwidth (subchannels) in the R1 and R3 zones, unlike traditional FFR, where only one-third of the bandwidth in the R3 zone is used. This is the reason that SRA-Case1 and SRA-Case3 achieve the highest levels of spectral efficiency in Figure 5.11. The advantage of using the SRA algorithm can clearly be seen when evaluating average spectral efficiency (using Equation 4.25), as shown in Figure 5.12. The use of segment

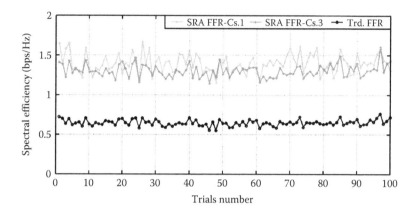

FIGURE 5.11
Spectral efficiency per traditional FFR and SRA FFR models.

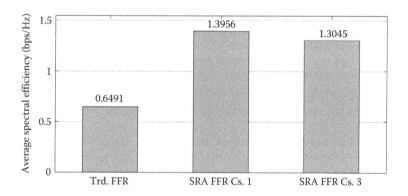

FIGURE 5.12
Average spectral efficiency per traditional FFR and SRA FFR models.

BC in any of the two models (SRA-Case1 or SRA-Case3) doubles spectral efficiency. This result refers to the amount of loss in spectral efficiency when the traditional FFR model is implemented without considering the utilization of segment BC.

5.2.2.6 Performance Comparison of Traditional FFR and SRA FFR

In this section, the performance evaluation of SRA algorithm is discussed. First, the responses of cases 1 and 3 are compared, and then the traditional FFR as an existing technology is used to evaluate the performance of SRA-Case1 and SRA-Case3, as follows.

Table 5.2 summarizes the output of segment BC in terms of cases 1 and 3. The number of served users, slot utilization, and data rate are averaged by the total number of trials. Segment BC can serve 30% and 45% of users from

TABLE 5.2

Performance Comparison between Cases 1 and 3

Case No.	Users	Slots	Data Rate (Mbps)	Average Subcarrier Efficiency (b/subcarrier/burst)
Case 1	6	30	1.5622	4.9056
Case 3	9	45	1.0003	2.1625

its full capacity (20 users) in cases 1 and 3, respectively. The reason for this is due to the availability of users in each case. However, following this logic, case 3 uses a greater number of slots (45 slots) than case 1, since serving more users means using more slots, as shown in Table 5.2. In contrast, the average data rate of case 1 exceeds that of case 3 (Figure 5.4), since the former can carry more data bits through its subcarriers than the latter, as shown in the last column of Table 5.2, where the average subcarrier efficiency is calculated. Subcarrier efficiency plays an important role in increasing the data rate. The value of subcarrier efficiency depends on the used modulation type, and the latter depends on user signal strength. Users near the base station have good signal strength, such as users in case 1. Therefore, the six users in case 1 carry more than twice the number of data bits than the nine users in case 3, as listed in Table 5.2.

In order to determine the advantages of the SRA algorithm, Table 5.3 shows the performance analysis of the traditional FFR, SRA-Case1, and SRA-Case3 algorithms.

The highest data rate, subcarrier efficiency, and spectral efficiency can be achieved when the SRA-Case1 model is applied, since segment BC in this model serves users near the base station and these users have a high SINR value, which leads to an increase in the values of the previous three metrics. In contrast, the highest number of served users and slot utilization can be achieved when the SRA-Case3 model is applied, since segment BC in this model serves more users and utilizes more slots than that in SRA-Case1. The percentage of served users and slot utilization are determined based on the DL subframe full capacity, which are 90 users and 390 slots. However, among the results as per Table 5.3, the arithmetic mean (see Section 4.5.3) of the average subcarrier efficiency in SRA-Case3 is slightly less than that in traditional FFR. This is due

TABLE 5.3

Performance Comparison between Traditional FFR, SRA-Case1 FFR, and SRA-Case3 FFR

Model Type	User %	Slot %	Data Rate (Mbps)	Mean Subcarrier Efficiency (b/subcarrier/burst)	Average Subcarrier Efficiency (bps/Hz)
Traditional FFR	77.7	74.35	7.049	2.3629	0.649
SRA-Case1	84.44	82.05	8.611	3.210	1.395
SRA-Case3	87.77	85.89	8.049	2.296	1.304

to the fact that segment BC in case 3 serves users far away from the base station with two possible types of MCS, such as 16 QAM with a 3/4 coding rate and 16 QAM with a 1/2 coding rate. These MCSs enable the subcarriers to carry a maximum and minimum number of data bits equal to 3 and 2, respectively.

The design of the SRA algorithm does not affect the normal operation of traditional FFR. On the contrary, the traditional FFR works properly, where it can serve the maximum number of users and utilize all the available slots. However, cases 3 and 4 serve more users than the other cases. Thus, the variation in population density was not considered in the design of SRA algorithm, since only one case should be chosen to represent segment BC.

In conclusion, two optimal solutions are suggested. If the target of the optimization is to increase the number of served users and resource utilization, then SRA-Case3 is the optimal solution. However, if the target of the optimization is to increase the data rate and spectral efficiency, then SRA-Case1 is the optimal solution. In order to choose either of them, it depends on the network administrator requirements. In general, SRA-Case1 and SRA-Case3 represent a real gain to enhance the performance of traditional FFR, as planned in phases 1 and 2 of the SRA algorithm (see Section 4.3.2).

5.3 Performance Analysis and Evaluation of the DRA FFR Algorithm

DRA FFR is the second algorithm that is proposed to enhance the performance of traditional FFR in terms of a variety of metrics, such as data rate, resource utilization, number of served users, and spectral efficiency, in addition to tackling the variation in population density. The DRA algorithm employs segment BC through four layers (A, B, C, and D) [4] in a dynamic manner, as planned in Section 4.4. In each attempt, the base station has the ability to serve the high population layer (winner layer) through segment BC. In order to show the advantages of the DRA algorithm, two types of user distribution are used: mobility pattern I, where the DRA is termed DRA-I, and mobility pattern II, where the DRA is termed DRA-II. It should be noted that there is no difference in the algorithm steps of DRA-I and DRA-II; they are the same as those of the DRA algorithm. The only difference is the method of users' distribution in the cell coverage area.

5.3.1 DRA-I FFR Algorithm Employing Mobility Pattern I

In mobility pattern I, the users are randomly distributed in the cell of interested. The process of random distribution of users is sequentially repeated frame by frame. In every attempt, the base station deals with different user locations for each DL subframe, thus giving a sense of mobility. In this section, the performance analysis of the layers in mobility pattern I is introduced first. The results of the DRA-I algorithm are then compared with those of the traditional FFR algorithm, as follows.

5.3.1.1 Per Layers in Mobility Pattern I

The DRA-I algorithm aims to assign resources dynamically into segment BC by serving the high population density layer. The analysis presented here discusses the performance (behavior) of the layers under the DRA-I algorithm.

1. *Based on data rate*: The data rate responses of the R1 zone, R3 zone, and segment BC (Seg. BC-I) are calculated by Equation 4.15 and depicted in Figure 5.13.

   Based on the proposed scenario, the R1 zone and segment BC can serve a maximum number of users equal to 60 and 20, respectively. The number of users who can be served in the R1 zone is triple that of segment BC, which is why the data rate level of segment BC is roughly one-third that of the R1 zone. The data rate of the R3 zone represents the aggregate data rate of 10 users located at the cell border. The cell border users have a low SINR value as a result of large path loss, which leads to the use of low MCS types. For the aforementioned reasons, the R3 zone shows the lowest data rate response in Figure 5.13.

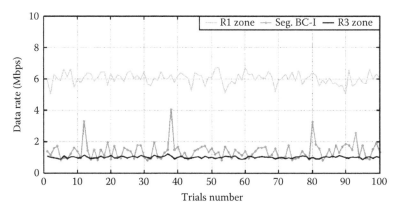

FIGURE 5.13
Data rate response of DL subframe parts in DRA-I.

2. *Based on subcarrier efficiency*: In order to get a closer picture of the data rate for the DL subframe parts, the subcarrier efficiencies for these parts are obtained based on Equation 4.20 and shown in Figure 5.14.

   The R1 zone carries more data bits than the other parts in the DL subframe, which is another reason for the increment of the data

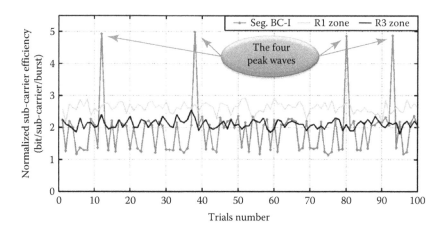

FIGURE 5.14
Subcarrier efficiency response of DL subframe parts in DRA-I.

rate in the R1 zone compared with other parts in Figure 5.13. The R1 zone serves users wherever they are in the cell center area, and these users enjoy eight types of MCS or, alternatively, eight types of burst profile (Table 4.3) from high to low MCS. This collection of MCSs enables the subcarriers of the R1 zone to carry more data bits than the other frame parts. In spite of the cell border users obtaining benefit from the FFR technique by increasing their SINR values, these values do not help them much when using the high MCS type, which limits the R3 zone response to around 2 bits, as shown in Figure 5.14. On the other hand, segment BC serves users wherever they are in the cell center area through the four layers with eight types of MCSs, which is the same as in the case of R1. Precisely, segment BC offers its services to the winner layer, which means that only two types of MCSs can be used at a time (per frame). In reference to the response of segment BC in Figure 5.14, the four high peak waves of data bits reflect the response of users who fall near the base station. These users occupy layer A, where layer A uses 64 QAM (5/6) and 64 QAM (3/4) MCS types, as listed in Table 4.3, which enable the subcarriers to carry 5 and 4.5 data bits, respectively. However, the average subcarrier efficiency (using Equation 4.21) of segment BC in Figure 5.15 has low performance compared with the R1 and R3 zones.

It makes sense that the output of segment BC in Figure 5.15 is the response of users located far away from the base station, since segment BC achieved less than 2 bits as a result of using low MCS types. The question, therefore, is which layer (A, B, C, or D) has been served by segment BC in DRA-I FFR? The answer to this question is given in the next section.

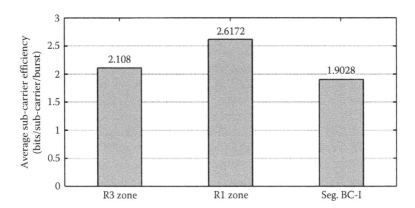

FIGURE 5.15
Average subcarrier efficiency of DL subframe parts in DRA-I.

3. *Based on users' distribution*: The concept of dynamic resource assignment used in the DRA algorithm depends on segment BC serving the layer of high population density. As mentioned in the first stage of the DRA algorithm (Section 4.4.2), the algorithm compares the number of users in each layer, and based on the comparison results, the crowded layer is served by segment BC. Figure 5.16 shows the winner layer in every trial, including the number of active users in it. It is clear from the figure that layers C and D are most frequently the winner layers. By contrast, layer A is rarely the winner, and layer B never had a chance to win the prize. The output of layer A confirms the results of Figure 5.14, where layer A wins the prize four times. The behaviors of layers have to do with the coverage area of each layer.

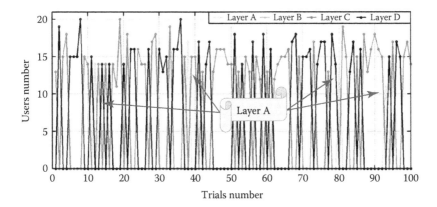

FIGURE 5.16
Winner layer mapping in DRA-I.

Statistically, the SINR thresholds of the four layers produce different coverage areas, since each layer is defined by upper and lower SINR thresholds (Table 4.4). A wide coverage area provides a great opportunity to accommodate more users and vice versa. To illustrate this ambiguity, Figure 5.17a shows the percentage area of each layer compared with the total cell center area. Layers C (31.5%) and D (32%) occupy the highest percentage (63.5%) of the cell center area, whereas layers A and B occupy 21% and 15.5%, respectively, of the cell center area. This means that layers C and D have a good chance of being the winner layer. The percentage area of each layer is calculated by dividing the area of the intended layer by the total cell center area. However, Figure 5.17b shows the users' distribution percentage in each layer, where 64% of users fall in layers C and D. Therefore, the response of segment BC is mostly as a result of serving users in layers C and D. The users' distribution percentage is calculated by dividing the number of users in the intended layer by the sum of the number of users in all layers for the entire test period. Finally, these findings illustrate the response of segment BC when considering the data rate and subcarrier efficiency, where the locations of layers C and D are far away from the base station. Moreover, these layers use low MCS types (Table 4.4) as a result of large path loss. The procedures that have been taken to select the appropriate MCS type are described in steps 22–41 of stage 3 in the DRA algorithm (see Section 4.4.2).

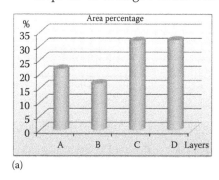
(a)

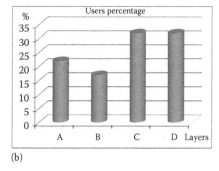
(b)

FIGURE 5.17
Statistical comparison between the area of service and the users' distribution. (a) Percentage area of each layer. (b) Users' distribution percentage in each layer.

4. *Based on DL subframe capacity*: The performance of segment BC has been examined in terms of data rate and subcarrier efficiency. However, the DRA algorithm aims to dynamically utilize segment BC in order to maximize resource utilization and the number of served users. Therefore, it is important to highlight the performance of segment BC in terms of the previous metrics. In reference to the proposed design (see Section 4.2.2), the R1 zone, R3 zone, and

segment BC can serve a maximum number of users equal to 60, 10, and 20, respectively. The R1 zone offers four slots to its users, while the R3 zone and segment BC offer five slots to their users. The numbers of active users and utilized slots are calculated by Equations 4.32 and 4.26, respectively, where the outputs of these equations are normalized by the trial number. Figure 5.18 shows the results of these equations.

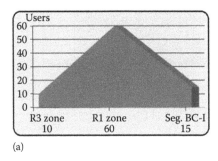

(a)

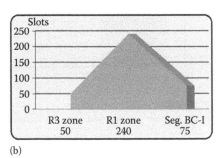

(b)

FIGURE 5.18
DL subframe capacity in DRA-I. (a) Average user number. (b) Average slot number.

Analyzing the results in Figure 5.18, the coverage areas of R1 and segment BC are similar and equal to the cell center area. Therefore, the R1 zone and segment BC can serve users as much as their capabilities allow, which means that they can utilize slots just as much. Bearing in mind that there are a lot of users in the cell center area where extra users should be served by segment BC (see step 7 in stage 1 of the DRA algorithm), the R1 zone works to its full capacity and utilizes all its slots. In contrast, and according to the random distribution of users, segment BC utilizes 75% (75/100) of resources from its full capacity (100 slots), since it follows a high population density. The policy is to serve the largest number of users, which is achieved through the first stage of the DRA algorithm. In addition, the number of users within the cell border always appears to be greater than 10. Thus, the R3 zone can serve the maximum number of users, as well as using the maximum number of slots.

5.3.1.2 *Performance Comparison between Traditional FFR and DRA-I FFR in Mobility Pattern I*

In this section, the results obtained from the DRA-I FFR algorithm are compared with those of the traditional FFR algorithm through various types of metrics, as follows:

1. *Based on DL subframe capacity:* An improvement achieved by the DRA-I algorithm is an increase in the number of served users and slot utilization, as depicted in Figure 5.19.

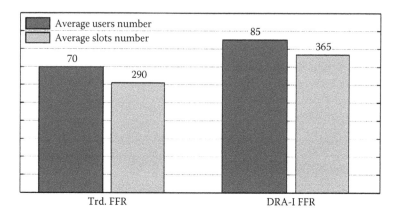

FIGURE 5.19
Average number of users and slots per traditional and DRA-I FFRs.

The average number of users (using Equation 4.36) in traditional FFR reaches its maximum value when 70 users are served in both the R1 and R3 zones. In turn, traditional FFR exploits the maximum number of slots (using Equation 4.30), where 290 slots are used in the R1 and R3 zones. On the other hand, using Equations 4.37 and 4.31, the DRA-I algorithm increases the number of served users to 85 and the number of utilized slots to 365, respectively. On the basis of this analysis, and although segment BC loses some users, the DRA-I FFR increases the number of served users and resource utilization compared with traditional FFR by 21.42% ((85−70)/70) and 25.86% ((365−290)/290), respectively, as a result of utilizing the resources in segment BC in a dynamic manner.

2. *Based on data rate*: The average data rates of the traditional FFR and DRA-I FFR, shown in Figure 5.20, can be obtained from Equations 4.18

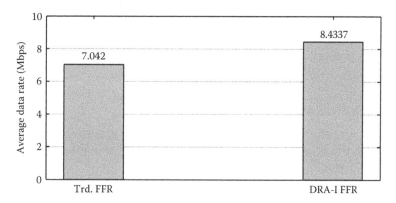

FIGURE 5.20
Average data rate of traditional and DRA-I FFRs.

and 4.19, respectively. As planned in stage 2 of the DRA algorithm (see Section 4.4.2), the DRA-I algorithm has the ability to increase the DL subframe capacity, where it utilizes more resources and serves more users as a result of using segment BC. Although segment BC in DRA-I serves users far away from the base station (layers C and D), and these users use low MCS types, DRA-I increases the average data rate by about 1.39 Mbps over that of traditional FFR. This data rate increase enhances the bandwidth utilization, where more data bits can be transmitted in a given bandwidth.

3. *Based on spectral efficiency*: The spectral efficiency of DRA-I FFR and traditional FFR algorithms are calculated using Equation 4.22. The delineation of spectral efficiency for the previously mentioned algorithms is illustrated in Figure 5.21.

Using all available subchannels in the R3 zone has a great impact in terms of increasing spectral efficiency, where a frequency reuse of 1 is achieved in the R3 zone, as analyzed in Section 4.5.4. Moreover, the DRA-I FFR algorithm increases the data rate to more than that of traditional FFR. Therefore, the average spectral efficiency (using Equation 4.25) shown in Figure 5.22 demonstrates the enhancement of DRA-I algorithm, where spectral efficiency is more than doubled.

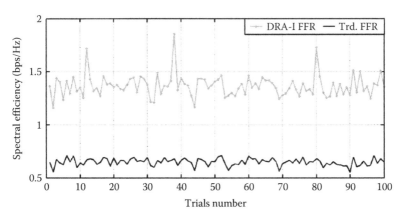

FIGURE 5.21
Spectral efficiency of traditional and DRA-I FFRs.

4. *Traditional FFR and DRA-I FFR*: Table 5.4 summarizes the results of DRA-I FFR and traditional FFR algorithms. The DRA-I algorithm serves a high population layer, which results in increasing the number of served users to 94.44% with respect to the maximum number of users in the DL subframe (90 users). This in return leads to an increase in resource utilization to 93.58% with respect to the full frame capacity (390 slots). The arithmetic mean of the subcarrier

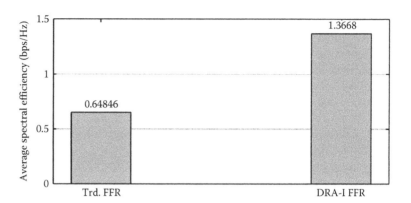

FIGURE 5.22
Average spectral efficiency per traditional and DRA-I FFRs.

TABLE 5.4

Performance Comparison between DRA-I and Traditional FFRs in Mobility Pattern I

Model Type	User %	Slot %	Data Rate (Mbps)	Mean Subcarrier Efficiency (b/subcarrier/burst)	Average Subcarrier Efficiency (bps/Hz)
Traditional FFR	77.7	74.35	7.042	2.3626	0.648
DRA-I	94.44	93.58	8.433	2.209	1.366

efficiency (see Section 4.5.3) is a little lower in DRA-I FFR than that in traditional FFR. This is because in every DL subframe, the DRA-I FFR serves the crowded layer and, according to the analysis presented in number 3 of Section 5.3.1.1, the most crowded layers are C and D (Figure 5.16). These layers serve the farthest districts in the cell center area. Thus, they use low modulation order as a result of large path loss, which in turn leads to fewer data bits in their subcarriers. However, what compensates for the losses in the number of data bits is an increase in the number of served users in segment BC. This proves that the validity of this conclusion is the increment of data rate that is achieved by DRA-I FFR (8.433 Mbps). In contrast, the utilization of all subchannels in the R3 zone increases bandwidth utilization, which results in an increase of the spectral efficiency of DRA-I FFR more than that of traditional FFR.

5.3.2 DRA-II FFR Algorithm Employing Mobility Pattern II

The DRA-I FFR algorithm under mobility pattern I enhances the performance of traditional FFR in different manners. The response of segment BC was mostly as a result of serving users in layers C and D, because these layers are the most congested. Therefore, it is important to use another mobility pattern to show the advantages of the DRA algorithm by employing all the layers (A, B, C, and D) to serve users. In mobility pattern II, it is assumed that a group of users is moving from the base station toward the border of the cell center area. Movement of this group starts from layer A and stops in layer D, as illustrated in Figure 5.23.

In each trial, a group of random users are distributed in the layer of interest. For instance, if layer A is under consideration, then a group of random users is distributed in layer A for a period of 25 trials. Since the test period is set to 100 trials, each layer hosts this group of users for 25 trials. The reason behind this type of mobility is to analyze the performance of the DRA algorithm in a fair way, where each layer has the ability to utilize the full capacity of segment BC for an equal time period (trial numbers). Referring to Figure 5.23, the highest performance level is achieved in layer A, whereas the lowest performance level is achieved in layer D. Therefore, using these layers for equal periods of time with full capacity gives the average response of these layers (segment BC). In the following sections, the performance analysis of the four layers is first presented, and then the results of DRA-II FFR are compared with those of traditional FFR.

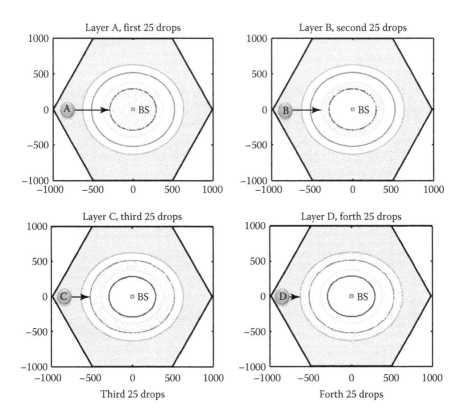

FIGURE 5.23
Users' distribution in DRA-II for 100 trials.

5.3.2.1 Per Layers in Mobility Pattern II

In this section, the behavior of the four layers in mobility pattern II is analyzed in terms of different metrics, as follows:

1. *Based on data rate*: The results of Equation 4.15 give the data rate of the DL subframe parts, as depicted in Figure 5.24. The responses of the R1 and R3 zones are the same as explained in number 1 of Section 5.3.1.1. Segment BC (Seg. BC-II) in mobility pattern II has a different response than that in mobility pattern I (Figure 5.13). Starting from layer A to layer D, each layer hosts users for a quarter of the duration of the test period. This leads to dividing the response of segment BC into four levels, as shown in Figure 5.24. Practically, the locations of the four layers from the base station are different; the layer near the base station uses high MCS as a result of low path loss, and the layer far away from the base station uses low MCS as a result of large path loss. Thus, the response of segment BC forms a descending ladder, since the area of service of segment BC moves from high to low MCS.

   In order to illustrate the benefits of using these segments, the average data rate of the DL subframe parts is illustrated in Figure 5.25.

   Given that, segment BC-II can serve one-third of the number of users in the R1 zone and twice the number of users in the R3 zone. Segment BC-II has the ability to produce a 47% (2.837/6.030) data rate compared with the R1 zone and 280.6% (2.837/1.011) compared with the R3 zone. This finding reveals the benefit of using segment BC in the DRA-II algorithm.

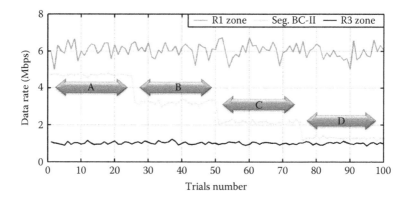

FIGURE 5.24
Data rate response of the DL subframe parts in DRA-II.

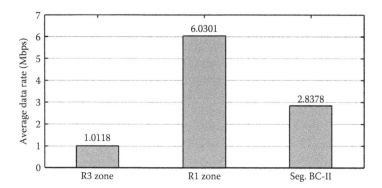

FIGURE 5.25
Average data rate response of the DL subframe parts in DRA-II.

2. *Based on subcarrier efficiency*: The data rate shown in Figure 5.25 is the result of using a collection of MCS types. In order to evaluate the response of each frame part, the subcarrier efficiencies of R1, R3, and segment BC-II are calculated by Equation 4.20 and illustrated in Figure 5.26. The responses of the R1 and R3 zones are the same as in number 2 of Section 5.3.1.1. The response of segment BC-II records four levels according to the four layers, where each level corresponds to a particular set of MCS types used by a specific layer. However, the DRA algorithm automatically tracks the densely populated regions in the cell center area and serves users in those regions with appropriate MCS types (see stage 3 of the DRA algorithm). Since four layers are used, this leads to the formation of three switching points where the burst profile is changed from one level to another.

The subcarrier efficiency of segment BC-II decreases with distance because the users are moving away from the base station toward the

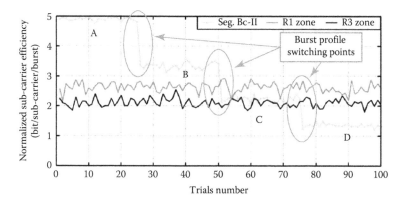

FIGURE 5.26
Subcarrier efficiency response of the DL subframe parts in DRA-II.

border of the cell center area (from high to low SINR). For instance, the subcarrier efficiency of the third level (layer C) is a result of using 16 QAM (3/4) and 16 QAM (1/2) modulations. Therefore, the subcarriers of users belonging to layer C should carry between 2 and 3 data bits. In order to show the amount of enhancement caused by segment BC, the average subcarrier efficiency of the DL subframe parts is calculated using Equation 4.21, as shown in Figure 5.27. Segment BC shows the highest subcarrier efficiency among the other parts; it represents 112.95% (2.956/2.617) of that in the R1 zone and 140.22% (2.956/2.108) of that in the R3 zone. These ratios indicate the amount of improvement that is caused by the DRA-II algorithm. In addition, a greater improvement in subcarrier efficiency can be achieved by segment BC if the target users are served in layer A for the whole test period, since layer A uses the highest MCS types in the system.

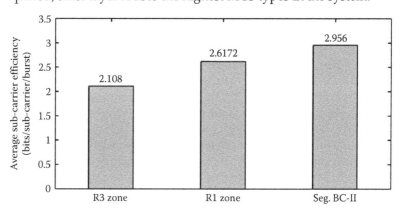

FIGURE 5.27
Average subcarrier efficiency per DL subframe part in DRA-II.

3. *Based on DL subframe capacity*: As previously mentioned when describing mobility pattern II (Figure 5.23), users are randomly distributed in each layer. The number of distributed users is equal to the maximum number of users that can be served by segment BC (20 users), to ensure that each layer has a chance to operate to its full load capacity. Figure 5.28 shows the number of users for the whole test period, where each layer hosts 20 users for 25 trials.

In each trial, 20 users are served by segment BC through the target layer. This is why the number of users in Figure 5.28 appears as a straight line with an amplitude of 20. This means that there is no wastage of resources in segment BC. Figure 5.29 shows the average number of served users and used slots, obtained by averaging the output of Equations 4.32 and 4.26, respectively, by the total number of trials. In reference to the proposed design (see Section 4.2.2), the number of slots reserved in the R1 zone is 240; the R3 zone, 50; and

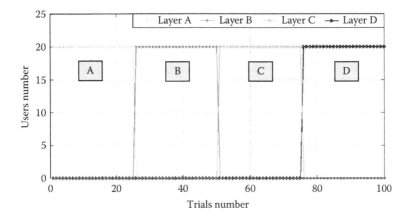

FIGURE 5.28
Number of users per layer in DRA-II.

segment BC, 100. Both the R1 and R3 zones work with full capacity, as well as segment BC. This is because segment BC in DRA-II exploits all its slots in the DL subframe, since it is able to serve the maximum number of users, as revealed in Figure 5.28.

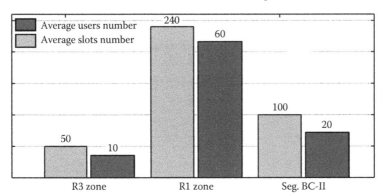

FIGURE 5.29
Average number of users and slots per DL subframe part in DRA-II.

5.3.2.2 Performance Comparison between Traditional FFR and DRA-II FFR in Mobility Pattern II

In order to show the benefits of the DRA algorithm in the second mobility, the results of DRA-II FFR are compared with those of traditional FFR in terms of a variety of metrics, as follows:

1. *Based on DL subframe capacity*: The average numbers of active users and utilized slots per traditional FFR and DRA-II FFR are depicted in Figure 5.30. The DRA-II is able to serve the maximum number

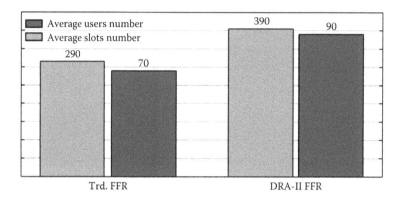

FIGURE 5.30
Average number of users and slots in traditional and DRA-II FFRs.

of users, which means it utilizes all the available slots in the DL
subframe. Since R1, R3, and segment BC are operating at full capacity, the DL subframe is fully utilized as shown in Figure 5.30. The
number of slots in DRA-II rises to 390 as a result of using the slots
reserved for segment BC. On the other hand, the number of served
users increases to 90 as a result of serving more users in segment
BC. The 390 slots and 90 users represent the full DL subframe
capacity. Therefore, in DRA-II FFR the resources are fully utilized
(100%), which makes DRA-II an important tool to address the lack of
resources in wireless communications.

2. *Based on data rate*: Figure 5.31 demonstrates the average data rate of
   traditional FFR and DRA-II FFR as a result of using Equations 4.18
   and 4.19, respectively.

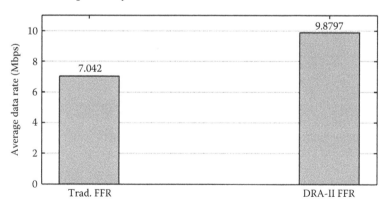

FIGURE 5.31
Average data rate in traditional and DRA-II FFRs.

It can be clearly shown that DRA-II FFR increases the data rate by about 2.8 Mbps compared with traditional FFR. This is due to the efficient resource utilization in segment BC, where all the layers are used evenly. Using all the layers evenly enables the base station to use eight types of MCSs, which results in an increase in the data rate in segment BC (Figure 5.25). Besides, the data rate of DRA-II is the average sum of 90 users, whereas the data rate of traditional FFR is the average sum of 70 users. For these reasons, DRA-II achieves a higher data rate than traditional FFR.

3. *Based on spectral efficiency*: The DRA-II FFR algorithm increases the data rate as a result of serving more users and utilizing more slots, where all the subchannels are used in the R3 zone. Therefore, this algorithm increases spectral efficiency. The spectral efficiencies (using Equation 4.22) of traditional FFR and DRA-II FFR are plotted in Figure 5.32.

In Figure 5.32, the output of traditional FFR swings between 0.5 and 0.7 (bps/Hz), since there is no change in the data rate level in the R1 or R3 zones. The output of DRA-II decreases slightly with time, due to the different levels of data rate produced by segment BC through the four layers (Figure 5.24). Every 25 trials, the DRA-II FFR line goes down to form a soft slope, as shown by the three points of turning levels in Figure 5.32. In spite of the output of DRA-II decreasing with time, it more than doubles spectral efficiency in comparison with traditional FFR, as revealed in Figure 5.33, where the average spectral efficiency is calculated by using Equation 4.25. Thus, the bandwidth utilization is improved, which is another reason to use the DRA-II algorithm in WiMAX network deployment.

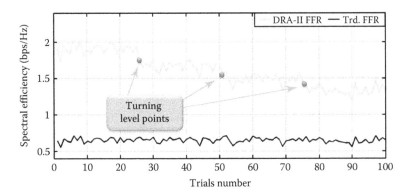

FIGURE 5.32
Spectral efficiency in traditional and DRA-II FFRs.

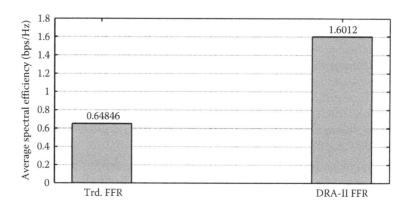

FIGURE 5.33
Average spectral efficiency of traditional and DRA-II FFRs.

TABLE 5.5

Performance Comparison between DRA-II and Traditional FFR in Mobility Pattern II

Model Type	User %	Slot %	Data Rate (Mbps)	Mean Subcarrier Efficiency (b/subcarrier/burst)	Average Subcarrier Efficiency (bps/Hz)
Traditional FFR	77.77	74.35	7.042	2.3626	0.648
DRA-II	100	100	9.879	2.560	1.601

4. *Traditional FFR and DRA-II FFR*: Table 5.5 summarizes the results obtained from DRA-II FFR and traditional FFR. The DL subframe is fully utilized, where segment BC serves users as much as its capacity allows, thus exploiting all the available slots. On the other hand, this finding confirms the losses in traditional FFR, where 25.65% (100%–74.35%) of resources are not used and 22.23% (100%–77.77%) of users are not served.

The arithmetic mean of subcarrier efficiency in Table 5.5 is increased as a result of employing the four layers in a fair way. Users near the base station can use high modulation order, which leads to an increase in the number of data bits in their subcarriers. In return, the data rate also increases, which causes an increase in spectral efficiency as well. The increase in the data rate in DRA-II FFR is a result of serving 90 users, and the increase in spectral efficiency is a result of increasing the data rate and the utilization of all the subchannels in the DL subframe. In DRA-II FFR, every subchannel in the R3 zone is used, unlike traditional FFR, where one-third of these subchannels are used.

5.4 Traditional FFR versus SRA FFR and DRA FFR

In this section, the results obtained for SRA FFR with its two models (cases 1 and 3), and the results obtained for DRA FFR with its two models (mobility patterns I and II), are compared with those for traditional FFR. Table 5.6 summarizes the results of the former models.

The results in Table 5.6 clearly show that SRA FFR and DRA FFR enhance the performance of traditional FFR in a variety of metrics, except the subcarrier efficiency. This metric fell slightly when it comes to SRA-Case3 and DRA-I, but it shows a good response when it comes to SRA-Case1 and DRA-II. The is because the service area of SRA-Case 3 is located far from the base station, and according to the modulation rules in Table 4.2 (see Section 4.3.1), case 3 uses low modulation and coding rate types, which causes the subcarriers to carry fewer data bits in SRA-Case3. The previous explanation can be repeated for DRA-I, since the most crowded layers are C and D. These layers are located far away from the base station, so they use low modulation order (Table 4.4).

Based on Table 5.6, the numbers of served users and utilized slots are increased dramatically when considering the SRA and DRA FFR models. DRA FFR is superior to SRA because the DRA works on the principle of serving high-population-density areas. Therefore, the DRA can serve more users and utilizes more slots, which increases its data rate more than that in SRA FFR and traditional FFR. However, the data rate in SRA-Case1 is higher than that in DRA-I since the former uses the highest set of modulation order, whereas the latter uses the lowest set of modulation order. The service area of case 1 is located next to the base station (users have high SINR), whereas the service areas of layers C and D in DRA-I FFR are located far from the base station (users have low SINR).

The increase of spectral efficiency in SRA and DRA FFRs is the result of using all the subchannels in the R3 zone. In SRA and DRA FFRs, a universal frequency reuse factor (FRF) is achieved where all the available bandwidth is utilized without breaking the conditions of traditional FFR. The DRA-II has the highest spectral efficiency compared with the other models since segment BC uses all the available modulation types evenly (eight types of burst profiles) and all the resources are utilized, which increases the data rate, thus leading to an increase in spectral efficiency.

TABLE 5.6

Performance Comparison Summary of All Proposed Models with Traditional FFR

Model Name		Served User %	Utilized Slot %	Data Rate (Mbps)	Mean Subcarrier Efficiency (b/subcarrier/burst)	Average Subcarrier Efficiency (bps/Hz)
Traditional FFR		77.7	74.35	7.049	2.3629	0.649
Static model	SRA-Case1 FFR	84.44	82.05	8.611	3.210	1.395
	SRA-Case3 FFR	87.77	85.89	8.0493	2.296	1.304
Dynamic model	DRA-I FFR	94.44	93.58	8.433	2.209	1.366
	DRA-II FFR	100	100	9.879	2.560	1.601

The results revealed by SRA and DRA FFRs show the design efficiency of these models in improving the performance of the traditional FFR. Table 5.7 shows the amount of reinforcement in SAR and DRA models when considering a grid of 19 cells for one DL subframe duration. The three metrics in Table 5.7 are calculated by multiplying the results obtained from the involved algorithms by the number of cells in the grid. It is assumed that all the base station in the grid enjoys the same conditions as the cell of interest and achieves the same amount of enhancement. In the other direction, Table 5.7 shows the amount of loss if SRA and DRA FFRs are not applied. However, the highest level of improvement is achieved in DRA-II, where 1710 users can be served in 19 cells per one DL subframe. In return, that means 380 (1710 – 1330) users are rejected in traditional FFR, which causes resource wastage of up to 1900 slots. In addition, losing this amount of resources leads to loss in data rate (53.77 Mbps), which reduces the spectral efficiency.

TABLE 5.7

Analysis of the Aggregate Improvement of 19 Cells for One DL Subframe

Model Name		Aggregate Served Users	Aggregate Utilized Slots	Aggregate Data Rate (Mbps)
Traditional FFR		1330	5510	133.931
Static model	Case 1 FFR	1444	6080	163.609
	Case 3 FFR	1501	6365	152.931
Dynamic model	DRA-I FFR	1615	6935	160.227
	DRA-II FFR	1710	7410	187.701

5.5 System Stability Evaluation per SRA FFR and DRA FFR Algorithms

The WiMAX Forum defines several sets of system parameters for IEEE 802.16e, such as 5, 10, and 20 MHz system bandwidth, to support a wide range of bandwidths to flexibly address the need for various spectrum allocation and the requirements of base stations [5]. So far, the SRA and DRA algorithms are executed using a 10 MHz system bandwidth. In this section, the SRA and DRA algorithms are implemented using different types of system bandwidth. These algorithms are carried out using 5 and 20 MHz system bandwidths. The benefit that can be obtained is that the response of the SRA and DRA FFR algorithms can be studied under different configurations. Moreover, the feasibility of these algorithms to work under these configurations can be determined. The related parameters regarding the 5 and 20 MHz bandwidths are discussed in Section 3.2 and listed in Table 3.1.

In addition, the network design and most of the base station parameters presented in Sections 4.2.1 and 4.2.2, respectively, are used to run the SRA and DRA algorithms at 5 and 20 MHz bandwidths. First, the 5 MHz bandwidth results are presented, followed by the 20 MHz bandwidth results.

5.5.1 SRA and DRA FFR System Stability Evaluation Using 5 MHz

In the 5 MHz WiMAX system profile [5], the numbers of data subcarriers and subchannels in the DL transmission direction are equal to 360 and 15, respectively. When the Partial Usage of Subchannels (PUSC) mode is used, these two parameters lead to form 120 slots in the R1 zone, 25 slots in the R3 zone, and 50 slots in segment BC. Knowing the number of slots required per user leads to specifying the number of users that can be served in each part of the DL subframe, such as the maximum number of users that can be served in the R1 zone is equal to 30; in the R3 zone, 5; and in segment BC, 10. The stability of the SRA and DRA algorithms is evaluated by using the 5 MHz system as follows.

5.5.1.1 SRA Stability Evaluation Using 5 MHz System Bandwidth

In this section, a variety of metrics are computed to explore the response of the SRA algorithm through SRA-Case1 and SRA-Case3, as follows:

1. *Based on DL subframe capacity*: The average numbers of utilized slots and active users are illustrated in Figure 5.34.

   Traditional FFR works properly according to its full capacity, where it exploits all the available slots (145) and serves the maximum number of users (35). The SRA-Case1 model serves 82.22% $((37/45) \times 100\%)$ of users, which leads to the exploitation of 79.48% $((155/195) \times 100\%)$ of the resources in the DL subframe. On the other

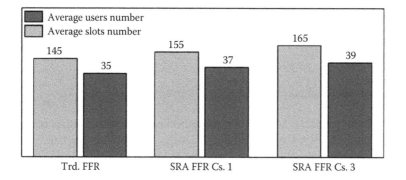

FIGURE 5.34
Average number of users and slots per traditional FFR, SRA-Case1, and SRA-Case3 at 5 MHz BW.

hand, the SRA-Case3 model serves 86.66% ($(39/45) \times 100\%$) of users, which leads to the exploitation of 84.61% ($(165/195) \times 100\%$) of the sources in the DL subframe. The responses of the SRA-Case1 and SRA-Case3 models in the 5 MHz system bandwidth are compatible with their responses in 10 MHz, as mentioned in Sections 5.2.2.1 and 5.2.2.2, where both models enhance the performance of traditional FFR. The number of active users and utilized slots in SRA-Case3 are greater than those in SRA-Case1, due to the available number of users in the service area of the former being greater than that in the latter.

2. *Based on data rate*: The average SRA-Case1 data rate exceeds those of SRA-Case3 and traditional FFR, as depicted in Figure 5.35. The SRA-Case1 and SRA-Case3 models increase the data rate of traditional FFR by about 17.51% ($(((4.1742 - 3.5521)/3.5521) \times 100\%))$ and 11.64% ($(((3.9659 - 3.5521)/3.5521) \times 100\%)$), respectively. The reason behind this enhancement is the utilization of additional resources in segment BC. The output of SRA-Case1 in Figure 5.35 gives a total data rate of 37 users, whereas the output of SRA-Case3 gives a total data rate of 39 users. In spite of the former serving fewer users than the latter, it has a higher data rate. The is because SRA-Case1 serves users near the base station through segment BC, and they enjoy good signal strength; therefore, they can use high modulation order. This conclusion is consistent with the trade-off study presented in Table 5.1.

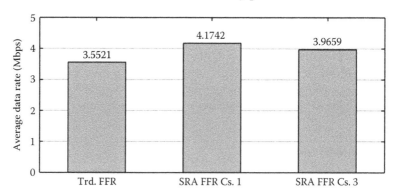

FIGURE 5.35
Average data rate of traditional FFR, SRA-Case1, and SRA-Case3 at 5 MHz BW.

3. *Based on spectral efficiency*: The obtained spectral efficiencies of traditional FFR, SRA-Case1, and SRA-Case3 are shown in Figure 5.36.

As spectral efficiency is directly proportional to the data rate, the system that can achieve the highest data rate should be able to achieve the highest spectral efficiency. Besides, the exploitation of all the available subchannels in the R3 zone leads to a greater increase

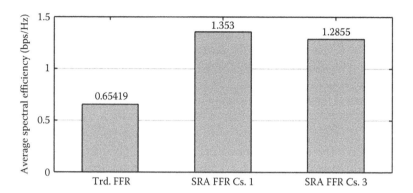

FIGURE 5.36
Average spectral efficiency per traditional FFR, SRA-Case1, and SRA-Case3 at 5 MHz BW.

in spectral efficiency in SRA-Case1 and SRA-Case3 than in traditional FFR. However, the behavior of the SRA-Case1 and SRA-Case3
algorithms at 5 MHz is identical to that in a 10 MHz system bandwidth (Table 5.3) in terms of increasing spectral efficiency. In these
two types of system bandwidth, spectral efficiency is doubled compared with that for traditional FFR.

5.5.1.2 DRA Stability Evaluation Using 5 MHz System Bandwidth

The stability of the DRA algorithm with its two mobility patterns (DRA-I and
DRA-II) is presented in this section and evaluated using the following metrics:

1. *Based on DL subframe capacity*: The average numbers of active users
   and used slots are illustrated in Figure 5.37. The DRA-I model
   serves 88.88% ($(40/45) \times 100\%$) of users, which results in using 87.17%
   ($(170/195) \times 100\%$) of the resources in the DL subframe. Similarly, the
   DRA-II model serves users as much as its capacity allows (100%),
   which results in using 100% of the resources in the DL subframe.

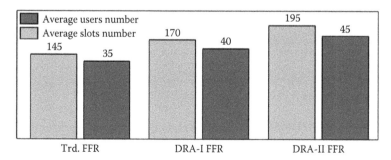

FIGURE 5.37
Average number of users and slots per traditional, DRA-I, and DRA-II FFRs at 5 MHz BW.

The is because all the resources (slots) in segment BC are used in mobility pattern II (DRA-II). However, the traditional FFR still maintains its level, whereby it exploits all the resources in the DL subframe. The behaviors of the DRA-I and DRA-II models at 5 MHz are the same as they are in the 10 MHz system bandwidth in terms of increasing the resource utilization and the number of served users, as illustrated in Figures 5.19 and 5.30 respectively.

2. *Based on data rate*: The data rates of traditional, DRA-I, and DRA-II FFR algorithms are illustrated in Figure 5.38. They form an ascending ladder as the result of the improvements caused by the DRA-I and DRA-II algorithms.

   The DRA-II model outperforms the DRA-I model, since the former can serve more users than the latter, as well as these users being evenly distributed in the cell center area. In other words, DRA-II serves users in all layers (A, B, C, and D), and these users experience different signal strengths (from high to low signal strength). Thus, eight types of modulation orders are used evenly, which causes an increase in the data rate of the DRA-II model. In contrast, DRA-I serves users in the crowded layer, which is concentrated in layers C and D. Layer C and D users suffer from large path loss, which reduces their signal strength. Therefore, they use a lower modulation order than layers A and B.

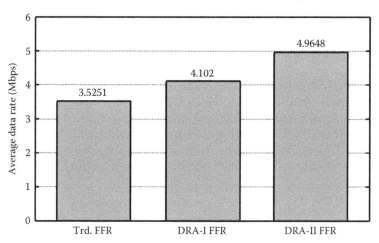

FIGURE 5.38
Average data rate per traditional, DRA-I, and DRA-II FFRs at 5 MHz BW.

3. *Based on spectral efficiency*: The spectral efficiency of traditional FFR and DRA FFR with its two types of mobility is shown in Figure 5.39. The results of spectral efficiency in Figure 5.39 are compatible with the data rate results in Figure 5.38, since the calculation of spectral efficiency depends on the aggregate cell data rate (see Equation 4.22).

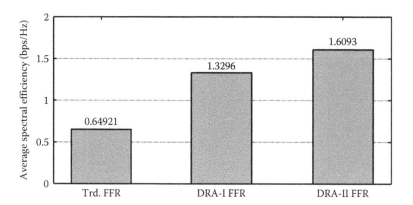

FIGURE 5.39
Average spectral efficiency of traditional, DRA-I, and DRA-II FFRs at 5 MHz BW.

The increase in spectral efficiency in the DRA-I and DRA-II models is the result of using all subchannels in the DL subframe, the same as in the 10 MHz system bandwidth (Table 5.6). In addition, DRA-II has higher spectral efficiency than DRA-I because the former achieves a higher data rate than the latter.

5.5.2 SRA and DRA FFR Systems' Stability Evaluation Using 20 MHz

In the 20 MHz WiMAX system profile, the number of data subcarriers and subchannels in the DL transmission direction is 1440 and 60, respectively [6]. According to the values of these two parameters and using the PUSC mode, the number of slots in the R1 zone equals 480; in the R3 zone, 100; and in segment BC, 200. Determining the number of slots needed by each user helps to calculate the maximum number of users that can be served in each part of the DL subframe. Consequently, the maximum number of users that can be served in the R1 zone is 120; in the R3 zone, 20; and in segment BC, 40. The stability of the SRA and DRA algorithms is evaluated by using a 20 MHz system bandwidth as follows.

5.5.2.1 SRA Stability Evaluation Using 20 MHz System Bandwidth

In this section, the performance of the SRA algorithm (including SRA-Case1 and SRA-Case3) is evaluated at 20 MHz system bandwidth by using the following metrics:

1. *Based on DL subframe capacity*: The average numbers of active users and utilized slots of traditional, SRA-Case1, and SRA-Case3 FFRs are depicted in Figure 5.40.

   Resource utilization is enhanced in the traditional FFR by both SRA-Case1 and SRA-Case3 FFRs, which is the same as in the 10

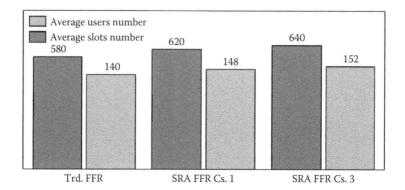

FIGURE 5.40
Average number of users and slots per traditional, SRA-Case1, and SRA-Case3 at 20 MHz BW.

MHz bandwidth (Table 5.3). This enhancement is a result of using the resources in segment BC, and utilizing more resources means serving more users. SRA-Case1 can serve 82.22% ($(148/180) \times 100\%$) of users, which leads to exploiting 79.48% ($(620/780) \times 100\%$) of the resources from its full capacity (180 users and 780 slots). In contrast, SRA-Case3 can serve 84.44% ($(152/180) \times 100\%$) of users, which leads to utilizing 82.05% ($(640/780) \times 100\%$) of resources SRA-Case3 can serve more users than SRA-Case1, which results in better resource utilization. On the other hand, traditional FFR can work to its full capacity, where it serves 140 users with the number of total slots equal to 580, as shown in Figure 5.40.

2. *Based on data rate*: The computation of average data rate per traditional FFR, SRA-Case1, and SRA-Case3 is illustrated in Figure 5.41.

The data rate in traditional FFR is improved as a result of serving more users in segment BC, where more resources are exploited by the

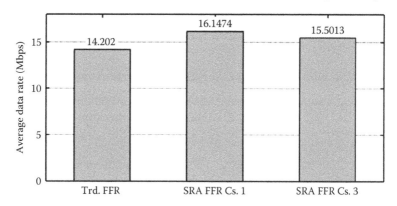

FIGURE 5.41
Average data rate per traditional, SRA-Case1, and SRA-Case3 FFRs at 20 MHz BW.

SRA-Case1 and SRA-Case3 algorithms. In spite of SRA-Case1 serving fewer users than SRA-Case3 (Figure 5.40), it shows a higher data rate than that of SRA-Case3. This is due to SRA-Case1 serving users near the base station with a high SINR value, which enables the base station to use the highest modulation order in the system. However, the different data rate values recorded by the SRA-Case1 and SRA-Case3 algorithms confirm the analysis presented (trade-off study) in Table 5.1 for these two cases in a 10 MHz system bandwidth.

3. *Based on spectral efficiency*: The average spectral efficiency of traditional FFR, SRA-Case1, and SRA-Case3 is shown in Figure 5.42.

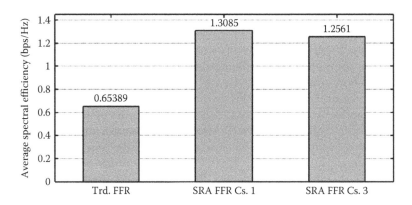

FIGURE 5.42
Average spectral efficiency in traditional, SRA-Case1, and SRA-Case3 FFRs at 20 MHz BW.

The outputs of SRA-Case1 and SRA-Case3 in Figure 5.42 enhance the performance of traditional FFR due to their using all the available subchannels in the DL subframe, unlike traditional FFR, where part of the subchannels is not used in the R3 zone. The SRA-Case1 FFR overcomes the SRA-Case3 FFR, since the former achieves a higher data rate than the latter. However, the result of spectral efficiency in SRA-Case1 and SRA-Case3 follows the same notation of enhancement in the 10 MHz system bandwidth, as presented in Figure 5.12. Therefore, these algorithms are able (reliable) to operate in the 20 MHz system bandwidth as well.

5.5.2.2 DRA Stability Evaluation Using 20 MHz System Bandwidth

In this section, the performance of the DRA FFR algorithm with two types of mobility patterns, DRA-I and DRA-II, is evaluated in a 20 MHz system bandwidth, as follows:

1. *Based on DL subframe capacity*: The average numbers of active users and utilized slots per traditional, DRA-I, and DRA-II FFRs are shown in Figure 5.43.

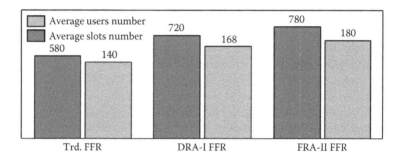

FIGURE 5.43
Average number of users and utilized slots in traditional, DRA-I, and DRA-II FFRs at 20 MHz BW.

In the 20 MHz system bandwidth, the number of slots is more than that in the 5 and 10 MHz bandwidths, which enables the system to serve more users. Both DRA-I and DRA-II FFRs enhance the performance of the traditional FFR, as revealed in Figure 5.43. Traditional FFR works properly where it can utilize the full capacity of the DL subframe, as in the 5 and 10 MHz system bandwidths. However, the DRA-I FFR can serve 93.33% ($(168/180) \times 100\%$) of users, which leads to exploiting 92.30% ($(720/780) \times 100\%$) of the resources from its full capacity (180 users and 780 slots). In contrast, DRA-II FFR utilizes all the available resources (100%) in the DL subframe since it can serve users as much as its capacity allows. There is no wastage of resources when DRA-II FFR is implemented, which is why DRA-II FFR performs better than DRA-I FFR.

2. *Based on data rate*: The average data rates of traditional, DRA-I, and DRA-II FFRs are illustrated in Figure 5.44.

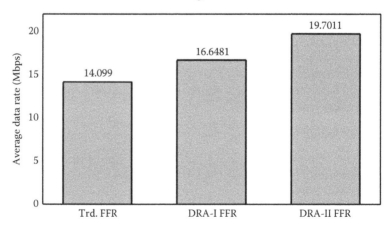

FIGURE 5.44
Average data rate in traditional, DRA-I, and DRA-II FFRs at 20 MHz BW.

The data rate of DRA-I FFR and DRA-II FFR improves the performance of traditional FFR at a 20 MHz system bandwidth at the same rate as it does at the 5 and 10 MHz bandwidths. The data rate in traditional FFR is increased by 18.08% $(((16.6481-14.099)/14.099)\times100\%)$ when DRA-I FFR is considered, while it is increased by 39.73% $(((19.7011-14.099)/14.099)\times100\%)$ when DRA-II FFR is applied. The enhancement of the data rate in the DRA-I and DRA-II algorithms is as a result of utilizing extra resources that were not used by traditional FFR. However, the increment of the data rate in DRA-II over DRA-I is related to the number of served users and slot utilization, since the former can serve more users and use more slots than the latter (Figure 5.43). In addition, DRA-II serves users uniformly distributed in the cell center area; however, users near the base station (layers A and B) can use a high modulation order, which results in an increase in the data rate.

3. *Based on spectral efficiency*: The average spectral efficiency of traditional, DRA-I, and DRA-II FFRs is illustrated in Figure 5.45. The results in Figure 5.45 show that DRA-I and DRA-II FFRs enhance the performance of traditional FFR. This enhancement is due to using all the available subchannels in the DL subframe, contrary to traditional FFR, where only one-third of the available subchannels in the R3 zone are used. DRA-II FFR achieves a high value of spectral efficiency compared with RDA-I FFR, since the former can provide a higher data rate than the latter (Figure 5.44). It is worth mentioning that the number of available slots at a 20 MHz bandwidth is much more than those available at a 5 or 10 MHz bandwidth; however, DRA-I and DRA-II show similar enhancement in spectral efficiency at 5 and 10 MHz bandwidths (Figures 5.39, 5.33, and 5.22, respectively). This finding proves that these algorithms can work properly at a 20 MHz bandwidth as well.

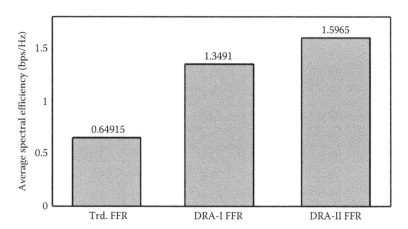

FIGURE 5.45
Average spectral efficiency in traditional, DRA-I, and DRA-II FFRs at 20 MHz BW.

5.5.3 System Stability Comparison in the SRA and DRA FFR Algorithms

Results obtained from analyzing the stability of the SRA and DRA FFR algorithms at a 5, 10, and 20 MHz system bandwidths are compared with those of the traditional FFR and presented in this section. The results are presented in Tables 5.8 through 5.10, respectively. In these tables, the average values of the selected metrics are considered when evaluating the stability performance of SRA-Case1, SRA-Case3, DRA-I, and DRA-II, including the results of traditional FFR.

Despite the disparity in the amount of improvement by all the proposed algorithms, Tables 5.8 through 5.10 reveal that these algorithms improved the performance of traditional FFR in different sizes of system bandwidth. In Sections 5.2 and 5.3, it has been extensively shown that the improvement of traditional FFR at a 10 MHz system bandwidth was caused by both SRA and DRA algorithms. The results of the SRA and DRA algorithms at 5 and 20 MHz system bandwidths, listed in Tables 5.8 and 5.10, respectively, follow the same notation of improvement at 10 MHz, since SRA-Case1, SRA-Case3, DRA-I, and DRA-II enhance the performance of traditional FFR by a variety of metrics. However, the amount of improvement in each system bandwidth is different according to the system parameters of the used bandwidth.

TABLE 5.8

Results' Comparison of 5 MHz Bandwidth

Metric Type		5 MHz				
		Traditional FFR	SRA-Case1	SRA-Case3	DRA-I	DRA-II
DL frame capacity	Active users	35	37	39	40	45
	Resource utilization	145	155	165	170	195
Data rate		3.5521	4.174	3.965	4.102	4.964
Spectral efficiency		0.6541	1.353	1.2855	1.329	1.6093

TABLE 5.9

Results' Comparison of 10 MHz Bandwidth

Metric Type		10 MHz				
		Traditional FFR	SRA-Case1	SRA-Case3	DRA-I	DRA-II
DL frame capacity	Active users	70	76	79	85	90
	Resource utilization	290	320	335	365	390
Data rate		7.049	8.6112	8.0493	8.433	9.8797
Spectral efficiency		0.6491	1.3956	1.3045	1.3668	1.6012

TABLE 5.10

Results' Comparison of 20 MHz Bandwidth

Metric Type		20 MHz				
		Traditional FFR	SRA-Case1	SRA-Case3	DRA-I	DRA-II
DL frame capacity	Active users	140	148	152	168	180
	Resource utilization	580	620	640	720	780
Data rate		14.202	16.1474	15.5013	16.648	19.7011
Spectral efficiency		0.6538	1.3085	1.2561	1.3491	1.5965

Therefore, it is helpful to show the results in a percentage format, as seen in Table 5.11.

In Table 5.11, two metrics are considered: active users and resource utilization. These metrics (users/slots) are calculated with respect to full DL subframe capacity, such as 45/195, 90/390, and 180/780 for a system bandwidth of 5, 10, and 20 MHz, respectively. Analyzing the results in Table 5.11, the DRA-II algorithm at a 5, 10, or 20 MHz system bandwidth can serve users as much as its capability allows, thus utilizing all the available resources in the DL subframe. The results of DRA-I are more than the results obtained from SRA-Case1 and SRA-Case3 at 5, 10, and 20 MHz system bandwidths. This is because the design of DRA-I is intended to serve areas that have a high population density, which enables the DRA-I algorithm to serve more users and use more resources than SRA-Case1 and

TABLE 5.11

Percentage Improvement in SRA and DRA Algorithms Using 5, 10, and 20 MHz System Bandwidths

Metric Type		Traditional FFR	SRA-Case1	SRA-Case3	DRA-I	DRA-II
5 MHz System Bandwidth						
DL frame capacity	Active user %	77.77	82.22	86.66	88.88	100
	Resource utilization %	74.35	79.48	84.61	87.17	100
10 MHz System Bandwidth						
DL frame capacity	Active user %	77.77	84.44	87.77	94.44	100
	Resource utilization %	74.35	82.05	85.89	93.58	100
20 MHz System Bandwidth						
DL frame capacity	Active user %	77.77	82.22	84.44	93.33	100
	Resource utilization %	74.35	79.48	82.05	92.30	100

SRA-Case3. However, the results of SRA-Case3 are greater than those of SRA-Case1 at a 5, 10, or 20 MHz system bandwidth due to the number of users in the service area in SRA-Case3 being more than that in SRA-Case1. Nevertheless, the data rate and spectral efficiency in SRA-Case1 are higher than those in SRA-Case3 (Tables 5.8 through 5.10). The results of SRA-Case1 and SRA-Case3 are conducive to a trade-off study; the option to use either of them depends on the network administrator requirements. Finally, the following can be concluded.

1. SRA and DRA algorithms enhance the performance of traditional FFR in different sizes of system bandwidth. Therefore, these algorithms are able to be used at 5, 10, and 20 MHz system bandwidths and can be used in cellular network deployment in order to increase the number of served users, resource utilization, data rate, and spectral efficiency.

2. The slight difference in the level of improvement shown by the SRA and DRA algorithms in the three types of system bandwidth in Table 5.11 is related to the number of slots reserved for the R1 zone, R3 zone, and segment BC in each type of system bandwidth. The ratio of reserved slots per zone or segment has the same value at a 5, 10, or 20 MHz system bandwidth, as follows:

    a. Consider the R1 zone at 5, 10, and 20 MHz system bandwidths. The ratio of slots reserved for the R1 zone equals 61.53%, since $\frac{120}{195} = \frac{240}{390} = \frac{480}{780} = 61.53\%$. The numbers 120, 240, and 480 represent the number of slots reserved for the R1 zone at 5, 10, and 20 MHz system bandwidth, respectively. However, 195, 390, and 780 represent the maximum number of slots in the DL subframe per 5, 10, and 20 MHz system bandwidth, respectively.

    b. Consider the R3 zone at 5, 10, and 20 MHz system bandwidths. The ratio of slots in the R3 zone equals 12.82%, since $\frac{25}{195} = \frac{50}{390} = \frac{100}{780} = 12.82\%$. The numbers 25, 50, and 100 represent the number of slots reserved for the R3 zone per 5, 10, and 20 MHz system bandwidth, respectively.

    c. Consider segment BC at 5, 10, and 20 MHz system bandwidths. The ratio of slots in segment BC equals 25.64%, since $\frac{50}{195} = \frac{100}{390} = \frac{200}{780} = 25.64\%$. The numbers 50, 100, and 200 represent the number of slots reserved for segment BC per 5, 10, and 20 MHz system bandwidth, respectively.

It is worth mentioning that 25.64% of the resources in segment BC has a significant impact in terms of improving the performance of traditional FFR in a variety of system parameters, such as at 5, 10, and 20 MHz bandwidths. Using 25.64% of resources in the proposed algorithms (SRA and DRA) leads to improvement of the performance of traditional FFR in terms of four metrics, as revealed in Tables 5.8 through 5.10.

5.5.4 Channel Capacity Evaluation Using 5, 10, and 20 MHz Bandwidth

The channel capacity of traditional FFR is computed using Equation 4.40, and the channel capacity of the SRA and DRA algorithms are computed using Equation 4.41. It should be noted that there is no difference in channel capacity calculation when SRA and DRA are applied; both show the same response. The reason for this is that from the standpoint of the used subchannels, both the SRA and DRA algorithms use the available subchannels in the R1 and R3 zones. Therefore, in this section, both the SRA and DRA algorithms are named as proposed FFRs and will be referred to as Pro. FFR. Following the same notation presented by [7], the channel capacities of the traditional FFR, FRF of 1, and proposed FFR against distance in the 10 MHz system bandwidth are plotted in Figure 5.46. The response of each of these models is as a result of user interaction with this base station. It is assumed that a user moves on the X-axis from the base station toward the cell border with a data load equal to the available subchannels in the R1 or R3 zones (see Section 4.5.6 for further explanation).

Four lines of channel capacity are plotted in Figure 5.46 as follows: Trd. FFR refers to traditional FFR when only the subchannels of segment A in the R3 zone are considered, Pro. FFR-BC refers to the proposed FFR when only the subchannels of segment BC in the R3 zone are considered, and Pro. FFR-ABC refers to the proposed FFR when all the subchannels of segments A, B, and C in the R3 zone are considered (which represents the performance of the SAR and DAR algorithms). Finally, the FRF of 1 refers to the frequency

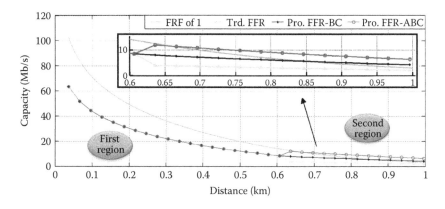

FIGURE 5.46
Channel capacity of traditional and proposed FFRs at 10 MHz BW.

reuse of 1 deployment scenario, where all the bandwidth is used by each cell in the grid or, in other words, without the FFR technique. The FRF of 1 is added to compare the proposed SRA and DRA algorithms with another type of deployment scenario to show the advantage of using the FFR technique, especially for cell border users.

The results shown in Figure 5.46 can be classified into two regions. The first region is from 36 to 635 m, and the second region is from 635 to 1000 m. In addition, when it comes to FFR channel capacity calculation, the first region represents the output of the R1 zone, whereas the second region represents the output of one or more segments in the R3 zone. The FRF of 1 shows the highest channel capacity compared with other models in the first region, since all the available effective bandwidth (EBW) is used in the DL subframe, where δ and ρ are equal to 1 (see Equation 4.39). In the FRF of 1, the DL subframe is used as one unit, and therefore there is no zoning or segmentation in the DL subframe. Besides, the SINR in the first region is high since users near the base station have high signal strength, while users far away from the base station suffer from low signal strength. Therefore, the FRF of 1 shows poor channel capacity in the second region of the cell coverage area (Figure 5.46). However, the output of the traditional and proposed FFRs in the first region is less than that of FRF of 1, which is related to the used bandwidth in the R1 zone. The R1 zone uses a share of the DL subframe bandwidth (EBW) that is as much as the number of orthogonal frequency-division multiplexing (OFDM) symbols reserved to the R1 zone in the DL subframe; this is not the case of FRF of 1, where all EBW is used. This is why the FRF of 1 achieves the highest channel capacity compared with the traditional and proposed FFRs in the first region. Moreover, the outputs of the traditional FFR, proposed FFR-BC, and proposed FFR-ABC are identical in the first region, since they use all the available subchannels in the R1 zone.

The results in the second region of Figure 5.46 are different. The SINR value decays exponentially with distance, since the SINR value of the intended user decreases when the user moves toward the cell border as a result of path loss. Therefore, all the channel capacity results decrease in the second region of Figure 5.46. This decrease in the channel capacity is due to a low SINR value and the number of used subchannels in each model. The channel capacity of traditional FFR decreases significantly in the second region, which is due to the loss of two-thirds of the available subchannels in the R3 zone. On the contrary, the channel capacity of the proposed FFR-ABC is increased in the second region. The reason is that all the subchannels are used in the R3 zone, which results in an increase in the channel capacity that is even more than that in FRF of 1. This fact shows the efficiency and effectiveness of the proposed design of SRA FFR and DRA FFR (represented by Pro. FFR-ABC). However, the channel capacity of the proposed FFR-BC in the second region of Figure 5.46 shows the advantage of using the subchannels of segment BC alone, which represents two-thirds of all subchannels in the

R3 zone. Moreover, it shows the disadvantages of traditional FFR when the subchannels of segment BC are not used.

The channel capacities for the FRF of 1, traditional FFR, proposed FFR-BC, and proposed FFR-ABC are depicted in Figures 5.47 and 5.48 for a system bandwidth of 5 and 20 MHz, respectively.

The results of the proposed FFR-ABC (Pro. FFR-ABC) at 5 and 20 MHz show the same type of enhancement as the traditional FFR (Trd. FFR) at a 10 MHz system bandwidth. The different level of enhancement achieved in the proposed FFR-ABC is related to the used system bandwidth and the available number of subchannels in the intended system bandwidth. For instance, the number of subchannels per segment BC in 20 MHz is equal to 40; in 10 MHz, 20; and in 5 MHz, 10. However, FFR-ABC triples the channel capacity

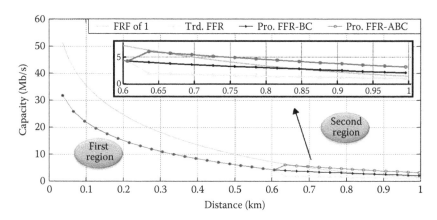

FIGURE 5.47
Channel capacity of traditional and proposed FFRs at 5 MHz BW.

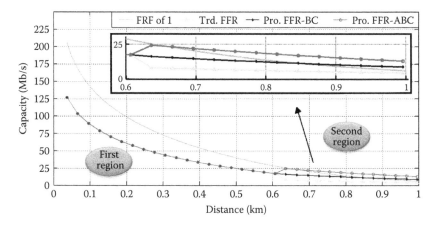

FIGURE 5.48
Channel capacity of traditional and proposed FFRs at 20 MHz BW.

in the R3 zone compared with traditional FFR at a 5, 10, and 20 MHz system bandwidth. The enhancement in channel capacity for traditional FFR is presented in Figures 5.46 through 5.48.

5.6 Results for SRA and DRA Algorithms

In this section, the performances of the SRA and DRA algorithms are compared with those of some other related works. The WiMAX 802.16e standard is a multiple access system (orthogonal frequency-division multiple access [OFDMA]) that is built on OFDM bases and can employ different slot definitions (permutation modes). In addition, different sets of system parameters can be used in the 802.16e standard, which leads to different scenarios in terms of allocating resources with various performances. The following is some of the research used for comparison.

Work done by [8] aims to enhance the resource utilization in FFR mobile WiMAX. A new model called load balance is introduced to increase the data rate in the base station. The load balance allocates resources based on the user channel conditions and the available free slots in the target zone or segments. The authors compare the results of the load balance model with two FFR methods named SINR based and distance based. Table 5.12 shows the comparison of the data rate achieved in this book (proposed FFR algorithms) with the three methods (load balance, SINR based, and distance based). In addition, the three FFR models presented by the authors are compatible with the traditional FFR model, which is explained in Section 2.5.

The results in Table 5.12 reveal that the SRA and DRA algorithms have increased the data rate of the base station more than the three methods

TABLE 5.12

Performance Comparison of SRA and DRA Models with Stiakogiannakis

Method	Model Name	Data Rate (Mbps)	Improvement in Load Balance (Mbps)	Improvement in SINR (Mbps)	Improvement in Distance (Mbps)
Proposed FFR algorithms	SRA-Case1	8.611	+0.311	+1.811	+2.011
	SRA-Case3	8.0493	−0.250	+1.249	+1.449
	DRA-I	8.433	+0.133	+1.633	+1.833
	DRA-II	9.879	+1.579	+3.079	+3.279
Stiakogiannakis et al. (2012) FFR algorithms	Load balance	8.3	—	—	—
	SINR	6.8	—	—	—
	Distance	6.6	—	—	—

presented by Stiakogiannakis. Although the data rate in SRA-Case3 shows a better result than that in the SINR and distance models, it records less megabits per second (0.2507) than the load balance model. SRA-Case3 serves users far away from the base station who suffer from large path loss, which reduces the ability to employ high modulation order, leading to a decline in the data rate. The last three fields in Table 5.12 show the amount of improvement in SRA and DAR against the load balance, SINR, and distance methods. Except for the SRA-Case3 model, all the proposed models achieve the highest data rate, which is due to efficiently exploiting the resources by the proposed algorithms.

The authors in [9] found the best system configuration parameters to increase the data rate of the FFR technique, by finding the suitable number of OFDM symbols for the R1 and R3 zones that give the highest data rate. In addition, different types of SINR threshold were examined to identify the effect of these thresholds on the data rate. The results of FFR (with different zone sizes and SINR thresholds) schemes were compared with the conventional FRF of 1 and FRF of 3 schemes. Table 5.13 shows the comparison with these schemes. For FFR, the highest data rates achieved by the authors are chosen for comparison. It should be noted that the FFR model presented by the authors is identical to traditional FFR. However, the three columns on the right side of Table 5.13 represent the amount of improvement achieved by each of the proposed models versus each of the authors' models. The FRFs of 1 and 3 are added to consider other deployment scenarios as a reference model for better comparison.

The results in Table 5.13 clearly show that the performance of the SRA and DRA algorithms is better than that obtained from the three schemes presented by Zhou and Zein. More slots in the R3 zone (segment BC) are utilized, thus leading to an increase in the data rate, unlike the traditional FFR, where only one-third of the slots in the R3 zone are used.

A new FFR was presented by [10] based on analyzing the frequency partitioning method in IEEE 802.20 and IEEE 802.16e FFR. In the new FFR, users

TABLE 5.13

Performance Comparison of SRA and DRA Models with Zhou and Zein

Method	Model Name	Data Rate (Mbps)	Improvement in FFR (Mbps)	Improvement in FR of 1 (Mbps)	Improvement in FFR Segment (Mbps)
Proposed FFR algorithms	SRA-Case1	8.611	+1.211	+3.131	+4.811
	SRA-Case3	8.0493	+0.649	+2.569	+4.249
	DRA-I	8.433	+1.033	+2.953	+4.633
	DRA-II	9.879	+2.479	+4.399	+6.079
Zhou and Zein (2008) algorithms	FFR	7.4	—	—	—
	FRF of 1	5.48	—	—	—
	FRF of 3 (segmentation)	3.8	—	—	—

TABLE 5.14

Performance Comparison of SRA and DRA Models with Han

Method	Model Name	Data Rate (Mbps)	Improvement in New FFR (Mbps)
	SRA-Case1	8.611	+0.451
Proposed FFR	SRA-Case3	8.0493	−0.110
algorithms	DRA-I	8.433	+0.273
	DRA-II	9.879	+1.719
Han et al. (2008) FFR algorithm	New FFR	8.16	—

in the inner region are able to use the subcarriers of the cell center area and cell border area, whereas users in the border region use the subcarrier of the cell border area. In order to control the interference between cells, the set of subcarriers in each cell border in the grid is different. The authors computed the data rate of new FFRs for different numbers of users, and upon the design of the new FFR, the DL subframe can serve a maximum number of users of 48. When the number of users reaches to the maximum value (48 users), the data rate is equal to 8.16 Mbps, which represents the full capacity of the new FFR. However, the SRA and DRA FFR algorithms achieved a higher data rate (Table 5.14) than that reported by Han et al. [10], except the data rate of SRA-Case3 (8.0493 Mbps), which it is little bit lower than 8.16 Mbps. Nevertheless, SRA-Case3 is able to serve 31 more users than the new FFR, where it served on average 79 users (Table 5.9). The SRA and DRA algorithms efficiently exploit the resources in the DL subframe, which lead to better performance.

Importantly, when comparing the data rate of DRA-II, which represents the full capacity of the DL subframe with the full capacity of the new FFR, the DRA-II FFR increases the data rate by about 1.719 Mbps compared with the new FFR, as shown in the last column of Table 5.14. Besides, DRA-II can serve almost double the number of users (90) than the new FFR (48). This is due to the fact that in DRA-II, all the available space in the DL subframe is used to serve more users, whereas in the new FFR, more partitioning is used to increase user signal strength. This finding emphasizes the importance of the use of the SRA and DRA algorithms in a cellular network.

5.7 Summary

Results related to the evaluation and analysis of the performance of the proposed SRA and DRA algorithms are presented in this chapter. The response of the SRA algorithm is analyzed through four cases, in order to find the best design parameters to maximize the performance of the SRA algorithm in terms of data rate, spectral efficiency, number of served users, and resource utilization. An analysis of these cases has led to a trade-off study, where the results proved that cases 1 and 3 are the best. These cases (1 and 3) give freedom to the network administrator to choose the most desirable system parameters that are appropriate for the needs of the network. The results demonstrated that case 1 is more effective for system needs in terms of increasing the data rate and spectral efficiency, whereas case 3 is more suitable for a system that requires serving more users and utilizing more resources.

The DRA algorithm is used to enhance the performance of the traditional FFR in addition to tackling the variation in population density. The results revealed that the DRA algorithm enhances the performance of the traditional FFR more than the SRA algorithm in terms of the same metrics that are used to evaluate SRA algorithm. The performance of the DRA algorithm is analyzed through two types of mobility patterns: DRA-I and DRA-II. The simulated results proved that when the resources in segment BC are evenly used (DRA-II), it has the highest performance and significantly enhances the performance of traditional FFR.

In addition, several procedures are employed to validate the results obtained from the SRA and DRA algorithms, such as comparing the results of these algorithms with existing technology (traditional FFR), implementing these algorithms by using different system parameters and bandwidth, using two types of FFR configuration (static and dynamic), and comparing the outcomes of these algorithms with other related works. Moreover, the channel capacity in SRA and DRA FFRs is enhanced three times compared with that of traditional FFR in the R3 zone.

Results show that the SRA and DRA algorithms are able to enhance the performance of traditional FFR in terms of six metrics: data rate, spectral efficiency, number of served users, resource utilization, channel capacity, and suitability for the WiMAX FFR cellular network deployment.

References

1. I. Akyildiz and X. Wang, *Wireless Mesh Networks*, vol. 3, John Wiley & Sons, Hoboken, NJ, 2009.
2. M. Salman, R. Ahmad, and A. Yahya, A new approach for efficient utilization of resources in WiMAX cellular networks, *Tehnicki vjesnik/Technical Gazette*, vol. 21, 1385-1393 2014.
3. M. Salman, R. Ahmad, and M. S. Al-Janabi, A new static resource and bandwidth utilization approach using WiMAX 802.16 e fractional frequency reuse base station, *Journal of Theoretical & Applied Information Technology*, vol. 70, 2014.

4. M. Salman, B. Ahmad, and A. Yahya, New dynamic resource utilization technique based on fractional frequency reuse, *Wireless Personal Communications*, vol. 83, pp. 1183–1202, 2015.

5. Forum, Mobile WiMAX—Part I: A technical overview and performance evaluation, 2006, p. 53.

6. J. G. Andrews, A. Ghosh, and R. Muhamed, *Fundamentals of WiMAX: Understanding Broadband Wireless Networking*, Pearson Education, London, 2007.

7. P. Godlewski, M. Maqbool, M. Coupechoux, and J.-M. Kélif, Analytical evaluation of various frequency reuse schemes in cellular OFDMA networks, in *Proceedings of the 3rd International Conference on Performance Evaluation Methodologies and Tools*, Greece, 2008, p. 32.

8. I. N. Stiakogiannakis, G. E. Athanasiadou, G. V. Tsoulos, and D. I. Kaklamani, Performance analysis of fractional frequency reuse for multi-cell WiMAX networks based on site-specific propagation modeling [wireless corner], *Antennas and Propagation Magazine, IEEE*, vol. 54, pp. 214–226, 2012.

9. Y. Zhou and N. Zein, Simulation study of fractional frequency reuse for mobile WiMAX, in *Vehicular Technology Conference, VTC Spring 2008*, Canada, 2008, pp. 2592–2595.

10. S. S. Han, P. Jongho, L. Tae-Jin, and H. G. Ahn, A new frequency partitioning and allocation of subcarriers for fractional frequency reuse in mobile communication systems, *IEICE Transactions on Communications*, vol. 91, pp. 2748–2751, 2008.

6

The Road Ahead

6.1 Introduction

WiMAX is a promising technology for high-speed data transmission to deliver a broadband service, and recently it has been the focus of attention by the research community. This chapter presents the conclusions and recommendations obtained from this book. Furthermore, it discusses some possible future research challenges that need to be studied in this area.

It is clearly mentioned in Chapter 1 that the main objective of this book is to investigate a novel approach to IEEE 802.16e fractional frequency reuse (FFR) WiMAX base stations that can enhance existing FFR algorithms. The focus of the present research was to provide different approaches to enhance not only the performance of FFR base stations in terms of several metrics, such as in most research studies, but also the preprocessing stages, such as designing a two-tier cellular network to analyze the DL subframe parts of the FFR technique in an interference environment and to make a decision on using these parts efficiently. Moreover, the designs of these preprocessing stages are supported by mathematical equations that show the stage design in mathematical formulas. The research work in this book focused on the following issues.

First, a comprehensive review of previous works on the FFR WiMAX base station was undertaken, and they were classified into two main categories based on the configuration type of the system parameters, namely, static and dynamic. Additionally, the resource and bandwidth utilization algorithms in the FFR technique were presented in six parts, and the advantages and disadvantages of these algorithms within each part were highlighted and discussed where possible. It was noted that most of the algorithms depend heavily on exploiting the existing DL subframe parts to increase resource exploitation. Moreover, during the process of searching for the deployment of WiMAX base stations in cellular networks, it was noted that the proposed algorithms mainly rely on using a fraction of the bandwidth to keep the interference at an acceptable level, thus resulting in inefficient utilization of resources and bandwidth. The FFR technique has been presented by the WiMAX Forum to increase the signal strength of cell border users, where part of the bandwidth and resources are not used to enable FFR to achieve its intended purpose, which makes FFR suitable for low-population environments, such as rural areas. These facts drove the researcher to make some contributions, such as the ones reported in this book.

Second, the basic concepts of the WiMAX technique were analyzed and discussed in this book, which encourages an understanding of the research subject. In general, these concepts provide a solid scientific base for analyzing and understanding WiMAX technology; in particular, it enables researchers to understand the concept of the FFR technique. These basic concepts represent the important features of WiMAX technology that have been used to create the new algorithms.

Third, in order to enable the FFR technique to work efficiently in urban areas to handle medium and high population densities, all the bandwidth (frequency reuse factor [FRF] of 1) needs to be used to increase the efficiency of FFR and make it suitable for crowded cities. In cellular network deployment, using all the bandwidth in each cell without affecting the channel quality of cell border users is a challenge. Therefore, a new algorithm, called static resource assignment (SRA) FFR, was proposed to overcome this challenge. In the SRA design, all the available bandwidth is used in each cell where an FRF of 1 is achieved without increasing the interference level between base stations. Four cases were evaluated for a trade-off study to identify the preferable system configuration parameters. SRA overcomes the existing FFR technique (traditional FFR) by a variety of metrics, such as resource utilization, number of served users, data rate, and spectral efficiency. However, choosing any of these cases depends on the network administrator requirements. The network administrator can easily change the parameters from one case to another without the need for additional devices or changing the network infrastructure. The SRA design is based on the static method, when the preferable case is chosen, and the base station then uses this configuration for a long period of time.

Fourth, as far as the objective of building up an efficient system for the FFR technique is concerned, another contribution has been conceived and named dynamic resource assignment (DRA) FFR. However, the developed SRA FFR algorithm was introduced by exploiting the benefits of the static configuration method. In static configuration, few control signals are required, since the system parameters are set earlier as a result of initial tests and kept unchanged for a specific period of time. However, it was noted that SRA FFR cannot adapt to the changes in user availability, since the number of users in an urban area or crowded city is subject to change according to several factors, such as the service area type (markets, schools, hospitals, universities, and government departments) and time of service (peak time). Peak time represents a burden on the base stations that provide services. These factors play an important role in determining the efficiency of the base stations. Consequently, the researcher of this book has developed a new algorithm called DRA FFR to tackle the variation in population density, which involves using the dynamic configuration method instead of the static one. The base station in DRA FFR is able to adapt the exploitation of resources according to the most crowded region in the cell center area. In order to avoid complexity, the base station serves densely populated areas by segment BC without changing the size of the zones. The adaption process leads to an increase in resource utilization, and at the same time, more users can be served. As a result, the data rate and spectral efficiency are increased as well. In order to show the advantages of the DRA FFR algorithm, two types of user distributions (mobility) were considered. These two types of mobility show the advantages of using the DRA FFR algorithm and its impact in terms of greatly enhancing the performance of traditional FFR.

Fifth, the stability and feasibility of the SRA and DRA FFR algorithms have been evaluated throughout this book. These algorithms have been implemented using three types of WiMAX system profiles: 5, 10, and 20 MHz system bandwidth. The results reveal that these algorithms can work properly in different system profiles and enhance the performance of traditional FFR through a variety of metrics, which make the SRA and DRA FFRs strong candidates for cellular network deployment. Moreover, the channel capacities for the aforementioned algorithms have been evaluated in different system bandwidths. The outcomes proved that the SRA and DRA FFRs have increased the channel capacity to more than threefold that of traditional FFR. This enhancement of channel capacity is related to using all the available bandwidth where a frequency reuse of 1 is achieved when deploying SRA and DRA FFRs in cellular networks.

The implementation of SRA and DRA FFR algorithms is in the media access control (MAC) layer. There is no need for more devices or changing the network infrastructure to run these algorithms. Thus, the implementation of these algorithms is inexpensive and uncomplicated; they could easily be adopted by the MAC layer of each base station in the grid.

Finally, the significance of these results and the impact that they have on the field of WiMAX wireless communications are summarized in three categories. First, WiMAX technology is a generation of modern and complicated wireless communications. The scientific materials and quantitative analysis presented in this book are a distinct starting point to understanding this technology. Moreover, the quantitative analysis deals with each part of the DL subframe of the FFR technique, which gives a clear and simple explanation to analyze the performance of these parts for any custom design. Second, important metrics were modeled and used to evaluate the performance of the FFR WiMAX base station. The modeling of these metrics was based on extensive mathematical analyses that meet the requirements of the FFR technique and involve the specifications of IEEE 802.16e WiMAX. The analyses of these metrics can be considered a base for further development and customization based on specific design requirements. Third, the new vision for the traditional FFR technique through SRA and DRA algorithms enables this technology to work in a crowded urban area at full capacity, which was not possible to achieve with traditional FFR. Therefore, applying these algorithms in WiMAX cellular networks fills an urgent need due to their ability to serve a larger number of users and meet the requirements of these users of resources.

6.2 Summary

In this book, two FFR algorithms were proposed for WiMAX cellular networks. The contribution of this book can be summarized as follows:

1. A WiMAX cellular network system was proposed and modeled. The modeling of the proposed system is based on analyzing the signal-to-interference-plus-noise ratio (SINR) level of the DL subframe parts of the FFR technique in an interference environment in order to enhance the performance of this technique.

2. The first algorithm, SRA FFR, combines efficient resource and bandwidth utilization and is used to improve the performance of existing traditional FFR. The SRA FFR is suitable for accommodating growing numbers of broadband wireless service users, since it enables the base station to exploit more bandwidth in the DL subframe and produces an FRF of 1 in each cell. The SRA algorithm structure is based on the static configuration method, which prevents this algorithm from adapting to changes in population density.

3. The second algorithm, DRA FFR, tackles the variation in population density, in addition to combining efficient resource and bandwidth utilization, where an FRF of 1 is achieved. The DRA algorithm structure is based on the dynamic configuration method, which enables the base station to serve more users and satisfy their demand for more resources, as is required for modern E-applications. In order to show the benefits of DRA FFR, two types of mobility patterns were proposed and modeled.

4. Selective metrics were analyzed, developed, and modeled to evaluate the performance of both the SRA and DRA algorithms. The evaluation is based on FFR configuration, coverage area, and mobility pattern in order to determine the optimal system parameters that maximize the performance of these algorithms.

5. From the results of SRA FFR, two optimal solutions were observed. If the optimization target is to increase the number of served users (87.11%) and enhance the resource utilization (85.89%), then case 3 is the optimal solution. In contrast, if the optimization target is to increase the data rate (up to 8.611 Mbps) and spectral efficiency

(1.395 bps/Hz), then case 1 is the optimal solution. On the other hand, in DRA FFR the following enhancements are achieved by the first/second mobility patterns: 94.44%/100% of users are served and 93.58%/100% of resources are utilized, where a data rate of up to 8.433/9.879 Mbps can be provided, which increases spectral efficiency to 1.366/1.601 bps/Hz. Moreover, the newly designed algorithms (SRA and DRA FFRs) increase the channel capacity three times in the R3 zone in comparison with traditional FFR.

6. Three types of WiMAX system profiles, 5, 10, and 20 MHz, were modeled and developed, as they are used in WiMAX base station deployment for different performances. The performances of the SRA and DRA algorithms were compared and evaluated using three types of system profiles in order to assess their feasibility and stability. The simulated results show that SRA and DRA enhance the performance of traditional FFR in terms of number of served users, resource utilization, data rate, and spectral efficiency.

Index

Milton Keynes UK
Ingram Content Group UK Ltd.
UKHW031133141024
449569UK00006B/232